STUDY GUIDE AND SOLUTIONS MANUAL

Introductory Chemistry

CONCEPTS & CONNECTIONS

SECOND EDITION

Charles H. Corwin

American River College

PRENTICE HALL Upper Saddle River, New Jersey 07458

S0-BBV-738

Senior Editor: *John Challice*
Associate Editor: *Mary P. Hornby*
Production Editor: *James Buckley*
Special Projects Manager: *Barbara A. Murray*
Supplement Cover Manager: *Paul Gourhan*
Production Coordinator: *Ben Smith*

© 1998 by Prentice-Hall, Inc.
Simon & Schuster / A Viacom Company
Upper Saddle River, NJ 07458

Printed in the United States of America

10 9 8 7 6 5 4 3 2 1

ISBN 0-13-908914-4

Prentice-Hall International (UK) Limited, *London*
Prentice-Hall of Australia Pty. Limited, *Sydney*
Prentice-Hall Canada, Inc., *Toronto*
Prentice-Hall Hispanoamericana, S.A., *Mexico*
Prentice-Hall of India Private Limited, *New Delhi*
Prentice-Hall of Japan, Inc., *Tokyo*
Simon & Schuster Asia Pte. Ltd., *Singapore*
Editora Prentice-Hall do Brasil, Ltda., *Rio de Janeiro*

Contents

Solutions to Odd-Numbered Exercises in the Textbook

Preface to the *Second Edition*

To the Student

It is my goal that this *Study Guide and Solutions Manual* will enable you to study chemistry more effectively while using the textbook **Introductory Chemistry: Concepts & Connections, 2/e**. This *Study Guide and Solutions Manual* is designed so that you can test your understanding of each topic and discover areas that require additional work. Thus, you will find self-test questions for each objective in the text.

There is also a computer-generated crossword puzzle following the self-test exercises to assist you in learning the key terms in a given chapter. The answers to the self-test questions as well as the crossword puzzle are included at the end of each chapter of this *Study Guide and Solutions Manual.*

The following guidelines will help you achieve maximum success in your chemistry course.

- **First**, take careful notes in lecture and pay close attention to the assignments given by your instructor.

- **Second**, read the assigned sections in the text and study the example exercises. Do the end-of-chapter exercises and check your answers in Appendix J of the text. If you cannot solve a problem or are having difficulty with an exercise, refer to the complete solution at the end of this study guide.

- **Third**, do the self-test exercises in this study guide to diagnose any lack of understanding in the assigned sections from the textbook. If you answer a question incorrectly, refer to the corresponding section in the textbook.

If you observe the above suggestions, this *Study Guide and Solutions Manual* will make your job of learning chemistry much more efficient. For those topics that still seem unclear, visit your instructor during office hours. Teachers enjoy helping students, especially those who are making a serious effort to do the assignments and learn the material.

Lastly, if you wish to share your chemistry experience, send me a note at the address below. I would like to hear about your success as well as any suggestions for improving these study aids.

Charles H. Corwin
Department of Chemistry
American River College
Sacramento, CA 95841

Introduction to Chemistry

Section 1.1 *Evolution of Chemistry*

1. What four elements composed everything in nature according to the beliefs of the ancient Greeks?
   - (a) air, earth, salt, and water
   - (b) air, ashes, fire, and water
   - (c) air, earth, fire, and water
   - (d) smoke, earth, fire, and water
   - (e) none of the above

2. Which of the following is a basic step in the scientific method?
   - (a) perform an experiment and collect data
   - (b) analyze experimental data and propose a hypothesis
   - (c) test a hypothesis and state a theory or law
   - (d) all of the above
   - (e) none of the above

Section 1.2 *Modern Chemistry*

3. In which of the following industries does chemistry play an important role?
   - (a) agriculture
   - (b) electronics
   - (c) paper
   - (d) transportation
   - (e) all of the above

4. Which of the following is *not* derived from a petrochemical?
   - (a) a gold coin
   - (b) a plastic toy
   - (c) an insecticide
   - (d) a synthetic fabric
   - (e) a blue dye

Section 1.3 *Learning Chemistry*

5. In a survey by the *American Chemical Society* of entering college students, which science course was considered the most relevant to their daily lives?
    (a) biology
    (b) chemistry
    (c) geology
    (d) physics
    (e) physiology

6. Which of the following is *not* a positive association with chemistry?
    (a) chemistry benefits society
    (b) chemistry has biomedical applications
    (c) chemistry involves dangerous experiments
    (d) chemistry topics are relevant to our daily lives
    (e) chemistry has many career opportunities

7. Connect the following nine dots using:

    (a) 5 straight, continuous lines.

    (b) 4 straight, continuous lines.

    (c) 3 straight, continuous lines.

Chapter 1 Key Terms

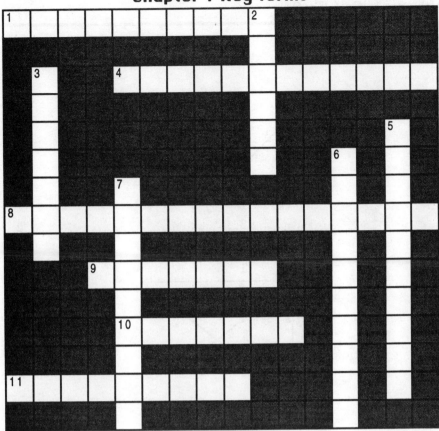

Across
1. a scientific procedure for collecting data
4. the study of biological substances and their reactions
8. a systematic investigation of nature (2 words)
9. the chemistry of substances containing carbon
10. a pseudoscience for changing lead into gold
11. the study of the composition of matter

Down
2. an extensively tested scientific proposal
3. the methodical exploration of nature
5. an initial explanation of experimental results
6. $P_1V_1 = P_2V_2$ (2 words)
7. substances not containing carbon

Chapter 1 Answers to Self–Test

1. c 2. d 3. e 4. a 5. b 6. c

7. (a)

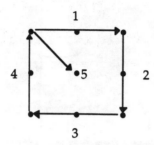

(b)

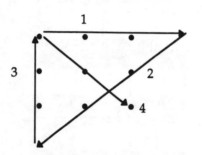

(c)

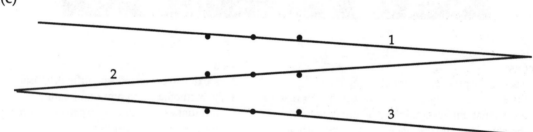

Chapter 1 Answers to Crossword Puzzle

Across
1. experiment
4. biochemistry
8. scientific method
9. organic
10. alchemy
11. chemistry

Down
2. theory
3. science
5. hypothesis
6. natural law
7. inorganic

Scientific Measurements

CHAPTER 2

Section 2.1 *Uncertainty in Measurements*

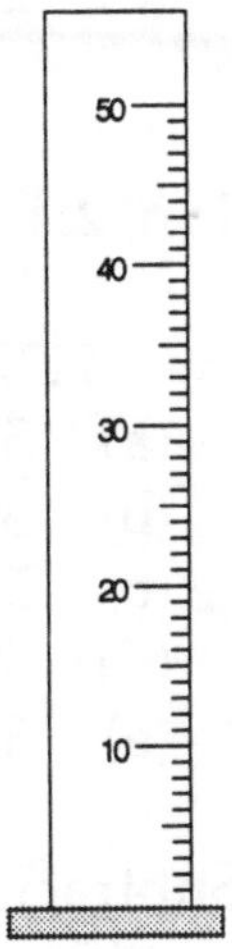

1. What instrument is shown in the diagram?
   (a) balance
   (b) graduated cylinder
   (c) metric ruler
   (d) volumetric pipet
   (e) volumetric flask

2. Which of the following quantities can be measured exactly?
   (a) length
   (b) mass
   (c) volume
   (d) time
   (e) no quantity can be measured exactly

Section 2.2 *Significant Digits*

3. How many significant digits are in the length measurement 20.05 meters?
   (a) 1
   (b) 2
   (c) 3
   (d) 4
   (e) none of the above

4. How many significant digits are in the mass measurement 0.0550 kilogram?
   (a) 2
   (b) 3
   (c) 4
   (d) 5
   (e) none of the above

Section 2.3 *Rounding Off Nonsignificant Digits*

5.   Round off the following measurement to three significant digits: 107.500 g.
     (a)   107 g
     (b)   108 g
     (c)   107.0 g
     (d)   107.4 g
     (e)   107.5 g

6.   Round off the measurement 2345 cm to three significant digits.
     (a)   230 cm
     (b)   234 cm
     (c)   235 cm
     (d)   2340 cm
     (e)   2350 cm

Section 2.4 *Adding and Subtracting Measurements*

7.   Add 7.77 g to 51.665 g and round off the sum to the proper significant digits.
     (a)   59.0 g
     (b)   59.4 g
     (c)   59.43 g
     (d)   59.44 g
     (e)   59.436 g

8.   Subtract 2.25 cm from 11.5 cm and round off the difference to the proper
     significant digits.
     (a)   9.0 cm
     (b)   9.2 cm
     (c)   9.20 cm
     (d)   9.25 cm
     (e)   9.3 cm

Section 2.5 *Multiplying and Dividing Measurements*

9.   Multiply 2.505 m times 1.75 m and round off the product to the proper number
     of significant digits.
     (a) $4.0\ m^2$
     (b) $4.00\ m^2$
     (c) $4.38\ m^2$
     (d) $4.384\ m^2$
     (e) $4.40\ m^2$

10. Divide 124.7 miles (mi) by 2.25 hours (h) and round off the quotient to the proper number of significant digits.
    (a) 50 mi/h
    (b) 55 mi/h
    (c) 55.0 mi/h
    (d) 55.4 mi/h
    (e) 55.5 mi/h

Section 2.6 *Exponential Numbers*

11. Express 0.000 000 000 000 000 000 000 000 000 000 010 as an exponential number.
    (a) 1.0×10^{-10}
    (b) 1.0×10^{-32}
    (c) 1.0×10^{-33}
    (d) 1.0×10^{32}
    (e) 1.0×10^{33}

12. Express the exponential number 5×10^{-9} as an ordinary number.
    (a) 0.000 09
    (b) 0.000 000 05
    (c) 0.000 000 005
    (d) 0.000 000 000 5
    (e) 5,000,000,000

Section 2.7 *Scientific Notation*

13. One gram of carbon contains 50,200,000,000,000,000,000,000 atoms. Express this number of carbon atoms in scientific notation.
    (a) 5.02×10^{21}
    (b) 5.02×10^{22}
    (c) 5.02×10^{23}
    (d) 50.2×10^{21}
    (e) 50.2×10^{22}

14. A carbon atom has a mass of 0.000 000 000 000 000 000 000 0199 g. Express this mass of a carbon atom in scientific notation.
    (a) 1.99×10^{-9} g
    (b) 1.99×10^{-21} g
    (c) 1.99×10^{-22} g
    (d) 1.99×10^{-23} g
    (e) 1.99×10^{-24} g

Section 2.8 *Unit Equations and Unit Factors*

15. Which of the following unit factors is derived from: 1 mile = 0.62 kilometer?
    (a) 1 mile/1 kilometer
    (b) 0.62 mile/1 kilometer
    (c) 0.62 kilometer/1 mile
    (d) 1 kilometer/0.62 mile
    (e) none of the above

16. How many significant digits are in the unit factor: 1 mile/0.62 kilometer?
    (a) 1
    (b) 2
    (c) 3
    (d) infinite
    (e) impossible to determine

17. Which of the following unit factors is derived from: 1 mile ≡ 1760 yards?
    (a) 1 mile/1 yard
    (b) 1760 miles/1 yard
    (c) 1760 yards /1 mile
    (d) 1760 yards/1760 miles
    (e) none of the above

18. How many significant digits are in the unit factor: 1 mile/1760 yards?
    (a) 1
    (b) 2
    (c) 3
    (d) infinite
    (e) impossible to determine

Section 2.9 *Unit Analysis Problem Solving*

19. After carefully reading a problem, what is the *first step* in the unit analysis method of problem solving?
    (a) write down the units asked for in the answer
    (b) write down the given value related to the answer
    (c) write one or more unit equations
    (d) write the two unit factors associated with each unit equation
    (e) apply a unit factor to convert the given value to the units of the answer

20. After carefully reading a problem, what is the *second step* in the unit analysis method of problem solving?
    (a) write down the units asked for in the answer
    (b) write down the given value related to the answer
    (c) write one or more unit equations
    (d) write the two unit factors associated with each unit equation
    (e) apply a unit factor to convert the given value to the units of the answer

21. After carefully reading a problem, what is the *third step* in the unit analysis method of problem solving?
    (a) write down the units asked for in the answer
    (b) write down the given value related to the answer
    (c) apply a unit factor to convert the given value to the units of the answer
    (d) enter numbers into your calculator
    (e) none of the above

Section 2.10 *The Percent Concept*

22. A 1¢ coin minted before 1982 is a metal alloy of copper and zinc. If the penny contains 2.869 g copper metal and 0.151 g zinc metal, what is the percentage of zinc in the coin?
    (a) 5.00%
    (b) 5.27%
    (c) 15.1%
    (d) 20.0%
    (e) 95.0%

23. A 10¢ coin minted after 1971 is a metal alloy of copper and nickel. If the 10¢ coin has a mass of 2.55 g and is 75.0% copper, what is the mass of nickel in the coin?
    (a) 0.102 g
    (b) 0.340 g
    (c) 0.638 g
    (d) 0.850 g
    (e) 1.91 g

Unit Analysis Problem Solving: Show all work for each of the following.

24. If the Sun is 93,000,000 miles from Earth, how many kilometers is the distance? (Given: One mile corresponds to a metric value of 1.61 kilometers.

    Unit Equation:

    Two Unit Factors:

    Unit Analysis Solution:

25. If the Sun is 93,000,000 miles from Earth, how long does it take for sunlight to reach the Earth? (Given: The velocity of light is 186,000 miles per second.)

    Unit Equation:

    Two Unit Factors:

    Unit Analysis Solution:

Chapter 2 Key Terms

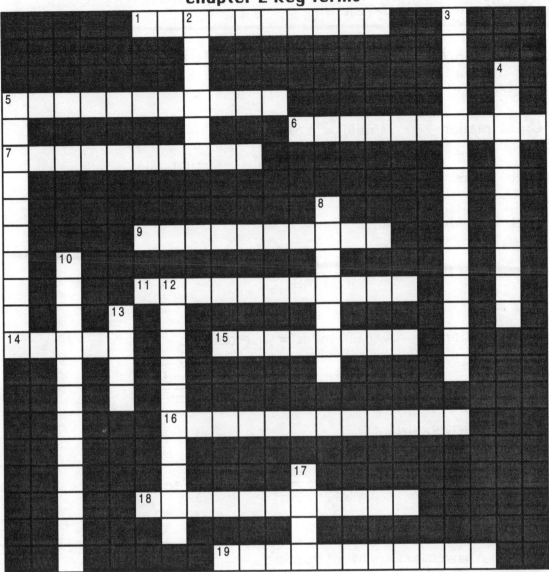

Across
1. 10^n (3 words)
5. eliminating nonsignificant digits (2 words)
6. the type of notation 3.35×10^7
7. a metric unit of length
9. of equal value
11. the digits obtained from a measurement
14. a metric unit of volume
15. a superscript number
16. a systematic problem solving method (2 words)
18. the inexactness of a measurement
19. a numerical value followed by units

Down
2. the force of gravity on an object
3. the digits often displayed in a calculator
4. a ratio of two equivalent quantities (2 words)
5. the relationship of a fraction and its inverse
8. parts per hundred parts
10. a statement of two equal values (2 words)
12. a device for obtaining a measurement
13. a metric unit of mass
17. the quantity of matter in an object

Chapter 2 Answers to Self–Test

1. b 2. e 3. d 4. b 5. b 6. e 7. d 8. e 9. c 10. d 11. b 12. c 13. b 14. d 15. c 16. b 17. c 18. d 19. a 20. b 21. c 22. a 23. c

24. **Unit Equation:** 1 mile = 1.61 kilometers

    Unit Factors: $\dfrac{1 \text{ mile}}{1.61 \text{ kilometers}}$ and $\dfrac{1.61 \text{ kilometers}}{1 \text{ mile}}$

    Unit Analysis Solution:

    $$9.3 \times 10^7 \text{ \sout{miles}} \times \frac{1.61 \text{ kilometers}}{1 \text{ \sout{mile}}} = 1.5 \times 10^8 \text{ kilometers}$$

25. **Unit Equation:** 186,000 miles = 1 second

    Unit Factors: $\dfrac{186\,000 \text{ miles}}{1 \text{ second}}$ and $\dfrac{1 \text{ second}}{186\,000 \text{ miles}}$

    Unit Analysis Solution:

    $$9.3 \times 10^7 \text{ \sout{miles}} \times \frac{1 \text{ second}}{186\,000 \text{ \sout{miles}}} = 5.0 \times 10^2 \text{ seconds}$$

Chapter 2 Answers to Crossword Puzzle

Across
1. power of ten
5. rounding off
6. scientific
7. centimeter
9. equivalent
11. significant
14. liter
15. exponent
16. unit analysis
18. uncertainty
19. measurement

Down
2. weight
3. nonsignificant
4. unit factor
5. reciprocal
8. percent
10. unit equation
12. instrument
13. gram
17. mass

The Metric System

Section 3.1 *Basic Units and Symbols*

1. Which of the following are the basic units and symbols for the metric system?
   - (a) centimeter (cm), gram (g), liter (L)
   - (b) centimeter (cm), gram (g), milliliter (mL)
   - (c) meter (m), gram (g), liter (L)
   - (d) meter (m), gram (g), liter (l)
   - (e) meter (m), gram (gm), liter (L)

2. What is the symbol for the metric unit microliter?
   - (a) cL
   - (b) mL
   - (c) ML
   - (d) μL
   - (e) none of the above

3. According to the metric system, 1 m = 10 _____.
   - (a) cm
   - (b) dm
   - (c) m m
   - (d) n m
   - (e) μm

4. According to the metric system, 1 g = 100 _____.
   - (a) kg
   - (b) dg
   - (c) cg
   - (d) mg
   - (e) μg

Section 3.2 *Metric Conversion Factors*

5. Which of the following is a unit factor corresponding to the unit equation: 1 m = 39.4 in.?
   (a) 1 m/1 in.
   (b) 1 m/39.4 in.
   (c) 39.4 in./39.4 m
   (d) 1 in./39.4 m
   (e) none of the above

6. Which of the following is a unit factor corresponding to the unit equation: 1 m ≡ 100 cm?
   (a) 1 m/1 cm
   (b) 100 m/100 cm
   (c) 1 cm/100 m
   (d) 100 cm/1 m
   (e) none of the above

Section 3.3 *Metric–Metric Conversions*

7. What is the three–step sequence, in order, for the unit analysis method of problem solving?
   (a) 1–relevant known value, 2–unknown units, 3–unit factor
   (b) 1–unknown units, 2–unit factor, 3–relevant known value
   (c) 1–unknown units, 2–relevant known value, 3–unit factor
   (d) 1–unit factor, 2–unknown units, 3–relevant known value
   (e) 1–unit factor, 2–relevant known value, 3–unknown units

8. What is the length in meters of a 20.0 mL test tube that measures 15.0 cm?
   (a) 0.150 m
   (b) 0.200 m
   (c) 1.50 m
   (d) 1500 m
   (e) 2000 m

9. The radius of a tin atom is 1.41×10^{-8} cm. What is the radius of the atom in micrometers?
   (a) 1.41×10^{-16} μm
   (b) 1.41×10^{-10} μm
   (c) 1.41×10^{-4} μm
   (d) 1.41×10^{-1} μm
   (e) 1.41×10^{5} μm

Section 3.4 *Metric–English Conversions*

10. Which of the following English–metric equivalents is correct?
    (a) 1 in. = 2.54 cm
    (b) 1 lb = 454 g
    (c) 1 qt = 946 mL
    (d) all of the above
    (e) none of the above

11. A stainless steel cylinder weighs 0.892 lb. What is the mass in kilograms?
    (a) 0.405 kg
    (b) 4.05 kg
    (c) 40.5 kg
    (d) 4050 kg
    (e) 405,000 kg

12. A glass bottle has a volume of 0.750 L. What is the volume in quarts?
    (a) 0.605 qt
    (b) 0.750 qt
    (c) 0.793 qt
    (d) 1.26 qt
    (e) 1.65 qt

Section 3.5 *Volume by Calculation*

13. A piece of gold foil has a volume of 0.645 mL. If the piece of gold measures 10.0 cm by 12.5 cm, what is the thickness of the foil?
    (a) 0.000516 cm
    (b) 0.00516 cm
    (c) 0.0516 cm
    (d) 0.516 cm
    (e) 0.806 cm

14. If an economy car has an 115 in.^{3} engine, what is the volume of the engine in cubic centimeters?
    (a) 1.88 cm^3
    (b) 7.01 cm^3
    (c) 292 cm^3
    (d) 742 cm^3
    (e) 1880 cm^3

Section 3.6 *Volume by Displacement*

15. What is the term for the method of determining volume by measuring the increase in water level after an object is completely immersed?
    (a) volume by displacement
    (b) volume by difference
    (c) hydrostatic method
    (d) density method
    (e) none of the above

16. A 200.0 g sample of cobalt metal is dropped into a 100-mL graduated cylinder containing 45.5 mL of water. If the resulting water level is 68.0 mL, what is the volume of the cobalt?
    (a) 12.5 mL
    (b) 22.5 mL
    (c) 32.0 mL
    (d) 54.5 mL
    (e) 113.5 mL

Section 3.7 *The Density Concept*

17. If the density of cobalt metal is 4.51 g/mL, which of the following is an associated unit factor?
    (a) 4.51 g/1 mL
    (b) 4.51 mL/1 g
    (c) 4.51 g/4.51 mL
    (d) 1 g/4.51 mL
    (e) 1 g/1 mL

18. If the density of ethyl alcohol is 0.789 g/mL, what is the volume of 35.5 g of ethyl alcohol?
    (a) 2.80 mL
    (b) 4.50 mL
    (c) 28.0 mL
    (d) 45.0 mL
    (e) 280 mL

19. A glass cylinder contains four separate liquid layers: mercury ($d = 13.6$ g/mL), chloroform ($d = 1.49$ g/mL), acetic acid ($d = 1.05$ g/mL), ether ($d = 0.708$ g/mL). If a marble ($d = 3.05$ g/mL) is added to the cylinder, where does it come to rest?
    (a) on top of the mercury layer
    (b) on top of the chloroform layer
    (c) on top of the acetic acid layer
    (d) on top of the ether layer
    (e) on the bottom of the cylinder

Section 3.8 *Temperature*

20. What is the freezing point and boiling point of water on the Celsius scale?
    (a) 0°C and 100°C
    (b) 0°C and 212°C
    (c) 32°C and 100°C
    (d) 32°C and 212°C
    (e) 273°C and 373°C

21. Given that liquid nitrogen boils at –196°C, what is the boiling point temperature on the Kelvin scale?
    (a) –469 K
    (b) –196 K
    (c) –77 K
    (d) 77 K
    (e) 196 K

Section 3.9 *Heat and Specific Heat*

22. Which of the following may express the total amount of heat energy in a sealed, insulated chamber?
    (a) 20.0°C
    (b) 68.0°F
    (c) 293.0 K
    (d) 20.0 kcal
    (e) all of the above

23. How many kilocalories of heat are required to raise 250.0 g of water from 20.0°C to 75.0°C?
    (a) 0.220 kcal
    (b) 4.55 kcal
    (c) 5.00 kcal
    (d) 13.8 kcal
    (e) 18.8 kcal

24. A 50.0 g sample of aluminum releases 420 calories when cooled from 100.0°C to 60.0°C. What is the specific heat of the metal?
    (a) 0.168 cal/g × °C
    (b) 0.210 cal/g × °C
    (c) 0.840 cal/g × °C
    (d) 4.76 cal/g × °C
    (e) 21.0 cal/g × °C

Unit Analysis Problem Solving: Show all work for each of the following.

25. If the radius of a nickel atom is 0.125 nm, what is the atomic radius of a nickel atom in picometers (pm)? (Given: $1 \text{ m} = 1 \times 10^{12}$ pm)

———————————

26. If a roll of nickel coins has a mass of 205 g, what is the mass of the roll of nickels in ounces? (Given: $1 \text{ lb} = 16$ ounces)

———————————

Chapter 3 Key Terms

Across

1. water boils at 100° on this temperature scale
5. a method for determining volume
7. the flow of energy from hot to cold particles
9. mass per unit volume
11. a problem solving method (2 words)
15. a volume equal to one milliliter (2 words)
19. a decimal system of measurement
20. 946 of these volume units equal 1 qt
21. 2.54 of these length units equal 1 in.
22. a statement of two equal quantities (2 words)
23. ice melts at 273 on this temperature scale

Down

2. the exact volume of 1 kg of water at 4°C
3. water freezes at 32° on this temperature scale
4. a mass equal to 1/454 pound
6. a unitless expression related to density (2 words)
8. a metric unit equal to one nutritional Calorie
10. the heat required to raise 1 g of water 1°C
11. the ratio of two equal quantities (2 words)
12. a system of measurement with seven base units
13. a mass approximately equal to two pounds
14. the value for water is 1.00 cal/g×°C (2 words)
16. a system of measurement with no basic units
17. the average energy of the particles in a system
18. a length approximately equal to a yard

Chapter 3 Answers to Self–Test

1. c 2. d 3. b 4. c 5. b 6. d 7. c 8. a 9. c 10. d 11. a 12. c 13. b 14. e 15. a
16. b 17. a 18. d 19. a 20. a 21. d 22. d 23. d 24. b

25. **Unit Analysis Solution:**

$$0.125 \,\cancel{nm} \times \frac{1 \,\cancel{m}}{1 \times 10^9 \,\cancel{nm}} \times \frac{1 \times 10^{12} \,pm}{1 \,\cancel{m}} = 125 \,pm$$

26. **Unit Analysis Solution:**

$$205 \,\cancel{g} \times \frac{1 \,\cancel{lb}}{454 \,\cancel{g}} \times \frac{16 \,oz}{1 \,\cancel{lb}} = 7.22 \,oz$$

Note: A summary of *Weights and Measures* including equivalent relationships is found in Appendix B of the textbook.

Chapter 3 Answers to Crossword Puzzle

Across
1. Celsius
5. displacement
7. heat
9. density
11. unit analysis
15. cubic centimeter
19. metric
20. mL
21. cm
22. unit equation
23. Kelvin

Down
2. liter
3. Fahrenheit
4. ram
6. specific gravity
8. kilocalorie
10. calorie
11. unit factor
12. SI
13. kilogram
14. specific heat
16. English
17. temperature
18. meter

CHAPTER

4

Matter and Energy

Section 4.1 *Physical States of Matter*

1. Which physical state demonstrates the greatest motion of individual particles?
   (a) solid state
   (b) liquid state
   (c) gaseous state
   (d) the motion of particles cannot be predicted for a physical state
   (e) the motion of particles is the same for each physical state

2. Which of the following describes a substance in the solid physical state?
   (a) The substance has a definite shape.
   (b) The substance has a fixed volume.
   (c) Particles have an ordered arrangement.
   (d) Particles vibrate in fixed positions.
   (e) all of the above

Section 4.2 *Elements, Compounds, and Mixtures*

3. Which of the following examples of matter can be separated into two or more substances by physical methods?
   (a) compound
   (b) element
   (c) heterogeneous mixture
   (d) pure substance
   (e) none of the above

4. Which of the following examples of matter must be separated into two or more substances by chemical methods?
   (a) compound
   (b) element
   (c) heterogeneous mixture
   (d) homogeneous mixture
   (e) none of the above

5. Sterling silver is an alloy of copper and silver with the percentage of copper always equal to 7.50%. Classify sterling silver as one of the following.
   (a) compound
   (b) element
   (c) heterogeneous mixture
   (d) homogeneous mixture
   (e) pure substance

Section 4.3 *Names and Symbols of the Elements*

6. Which of the following elements is not one of the ten most abundant in the Earth's crust, oceans, and atmosphere?
   (a) hydrogen
   (b) oxygen
   (c) silicon
   (d) titanium
   (e) uranium

Section 4.4 *Metals, Nonmetals, and Semimetals*

7. Which of the following is a typical physical property of a nonmetal?
   (a) brittle in the solid state
   (b) nonconductor of heat and electricity
   (c) low density and melting point
   (d) occurs as a solid or a gas at normal conditions
   (e) all of the above

8. Which of the following is a characteristic chemical property of a metal?
   (a) reacts by gaining electrons
   (b) reacts by losing electrons
   (c) reacts by sharing electrons
   (d) reacts with other metals
   (e) none of the above

9. Which of the following elements is an example of a semimetal?
   (a) H
   (b) C
   (c) Fe
   (d) Ge
   (e) Sn

10. Use a periodic table and predict which of the following elements is a gas at 25°C and 1 atmosphere pressure?
    (a) N
    (b) Na
    (c) Nb
    (d) Nd
    (e) none of the above

11. Use a periodic table and predict which of the following elements is a liquid at 25°C and 1 atmosphere pressure?
    (a) B
    (b) Ba
    (c) Be
    (d) Br
    (e) none of the above

12. Use a periodic table and predict which of the following elements is a solid at 25°C and 1 atmosphere pressure?
    (a) H
    (b) He
    (c) Hg
    (d) Ho
    (e) none of the above

Section 4.5 *Compounds and Chemical Formulas*

13. In which of the following compounds are the elements present in a definite proportion by mass?
    (a) H_2O
    (b) CO_2
    (c) N_2O_4
    (d) NaCl
    (e) all compounds have constant composition

14. Ammonium phosphate is used in fertilizer to replenish nitrogen to the soil. If the formula is $(NH_4)_3PO_4$, what is the total number of atoms in one molecule?
    (a) 13
    (b) 16
    (c) 18
    (d) 20
    (e) none of the above

Section 4.6 *Physical and Chemical Properties*

15. Which of the following is an example of a physical property?
    (a) color
    (b) crystalline form
    (c) melting point
    (d) physical state
    (e) all of the above

16. Which of the following is an example of a chemical property?
    (a) density
    (b) ductility
    (c) electrical conductivity
    (d) flammability
    (e) hardness

17. Which of the following is an example of a chemical property of methyl alcohol?
    (a) methyl alcohol and sulfuric acid yield methyl ether
    (b) methyl alcohol in animals produces blindness
    (c) methyl alcohol and sodium metal release hydrogen gas
    (d) methyl alcohol and formic acid give rum flavor
    (e) all of the above

Section 4.7 *Physical and Chemical Changes*

18. Which of the following observations is *not* evidence for a physical change?
    (a) condensation
    (b) crystallization
    (c) deposition
    (d) explosion
    (e) sublimation

19. Which of the following observations is *not* evidence for a chemical change?
    (a) producing bubbles after mixing solutions
    (b) giving a precipitate after mixing solutions
    (c) liberating heat after mixing solutions
    (d) changing color after mixing solutions
    (e) dissolving a solute in solution

Section 4.8 *Conservation of Mass*

20. If 558 kg of powdered iron react with powdered sulfur to produce 879 kg of iron sulfide, what is the mass of reacting sulfur?
    - (a) 321 kg
    - (b) 558 kg
    - (c) 879 kg
    - (d) 1437 kg
    - (e) impossible to predict from the given information

Section 4.9 *Potential and Kinetic Energy*

21. Which of the following is an example of kinetic energy?
    - (a) coliiding molecules
    - (b) accelerating roller coaster
    - (c) swinging in a swing
    - (d) rolling wheels
    - (e) all of the above

22. Which of the following physical states possesses the greatest attraction between individual particles?
    - (a) solid state
    - (b) liquid state
    - (c) gaseous state
    - (d) plasma state
    - (e) the attraction is the same for each physical state

23. According to the kinetic theory, temperature measures which of the following?
    - (a) molecular attraction
    - (b) molecular mass
    - (c) molecular motion
    - (d) molecular size
    - (e) none of the above

Section 4.10 *Conservation of Energy*

24. A nuclear power plant uses radioactive uranium to convert water to steam; steam then drives a turbine that turns a generator for electricity. What are the two forms of energy represented by uranium and the turbine?
    - (a) chemical and mechanical energy
    - (b) chemical and heat energy
    - (c) electrical and mechanical energy
    - (d) nuclear and heat energy
    - (e) nuclear and mechanical energy

25. According to the law of conservation of mass and energy, which of the following is true for a nuclear fission bomb?
   (a) the mass of the bomb and the fission products are identical
   (b) the energy of the bomb and the fission products are identical
   (c) a small amount of matter is converted into energy
   (d) all of the above
   (e) none of the above

Chemical Symbols of the Elements

26. Provide the chemical symbol for the name of the following elements.

   (a) boron –   (b) neon –

   (c) chlorine –   (d) cadmium –

   (e) krypton –   (f) cobalt –

   (g) nickel –   (h) xenon –

   (i) manganese –   (j) zinc –

27. Write the name of the element for each of the following chemical symbols.

   (a) Sn –   (b) Hg –

   (c) Fe –   (d) Cu –

   (e) Sb –   (f) Au –

   (g) K –   (h) Na –

   (i) Pb –   (j) Ag –

Chapter 4 Key Terms

Across
1. a change that does not alter composition
8. an element that reacts with metals
9. matter with constant composition
10. a shiny, high-density element
11. a property that involves another substance
15. a mixture with variable properties
18. it cannot be broken down by chemical reaction
20. the energy of a moving particle
21. a change of state from a gas to a liquid
22. the two related quantities in $E = mc^2$ (3 words)

Down
2. abbreviation for the name of a compound
3. all substances have this type of composition
4. a metalloid
5. a change of state from a liquid to a gas
6. a chemical reaction is a chemical _____
7. it cannot be created nor destroyed
9. a direct change of state from a solid to a gas
10. the property of being hammered into a foil
11. it can be broken down into elements
12. the stored energy of chemical composition
13. abbreviation for the name of a compound
14. the property of being drawn into a wire
15. a mixture with consistent properties
16. a physical or chemical characteristic
17. it can be solid, liquid, or gas
19. it can be nuclear, electrical, or heat

Chapter 4 Answers to Self–Test

1. c 2. e 3. c 4. a 5. d 6. e 7. e 8. b 9. d 10. a 11. d 12. d 13. e 14. d 15. e
16. d 17. e 18. d 19. e 20. a 21. e 22. a 23. c 24. e 25. c

26. Symbols of Chemical Elements

(a)	B	(b)	Ne
(c)	Cl	(d)	Cd
(e)	Kr	(f)	Co
(g)	Ni	(h)	Xe
(i)	Mn	(j)	Zn

27. Names of Chemical Elements

(a)	tin	(b)	mercury
(c)	iron	(d)	copper
(e)	antimony	(f)	gold
(g)	potassium	(h)	sodium
(i)	lead	(j)	silver

Chapter 4 Answers to Crossword Puzzle

Across
1. physical
8. nonmetal
9. substance
10. metal
11. chemical
15. heterogeneous
18. element
20. kinetic
21. condensation
22. mass and energy

Down
2. symbol
3. constant
4. semimetal
5. vaporization
6. change
7. mass
9. nonmetal
12. potential
13. formula
14. ductile
15. homogeneous
16. property
17. state
19. energy

Models of the Atom

Section 5.1 *Dalton Model of the Atom*

1. Which of Dalton's atomic theory proposals was later proven to be incorrect?
   - (a) An element is composed of tiny particles called atoms.
   - (b) All atoms of the same element have the same mass.
   - (c) Atoms of different elements combine to form compounds.
   - (d) Compounds contain atoms in small whole-number ratios.
   - (e) Atoms can combine in more than one whole-number ratio.

Section 5.2 *Thomson Model of the Atom*

2. What was the first experimental evidence for subatomic particles?
   - (a) The deflection of cathode rays by a magnetic field.
   - (b) The deflection of canal rays by an electric field.
   - (c) The lines in the emission spectrum of electrically excited atoms.
   - (d) The emission line splitting in the presence of a magnetic field.
   - (e) none of the above

3. Which subatomic particle is in the nucleus and has a relative charge of 1+?
   - (a) alpha
   - (b) electron
   - (c) neutron
   - (d) proton
   - (e) none of the above

Section 5.3 *Rutherford Model of the Atom*

4. What are the approximate diameters of an atom and its nucleus?
   - (a) 10^8 cm and 10^{-13} cm, respectively
   - (b) 10^{-8} cm and 10^{-13} cm, respectively
   - (c) 10^{-13} cm and 10^8 cm, respectively
   - (d) 10^{-13} cm and 10^{-8} cm, respectively
   - (e) 1 cm and 0.1 cm, respectively

5. State the subatomic particle having a relative charge of zero and an approximate mass of one atomic mass unit.
   (a) alpha
   (b) electron
   (c) neutron
   (d) positron
   (e) proton

Section 5.4 *Atomic Notation*

6. Using atomic notation, indicate the isotope having 10 p$^+$, 12 n^o, and 10 e$^-$.
   (a) $^{12}_{10}$Ne
   (b) $^{22}_{10}$Ne
   (c) $^{22}_{12}$Ne
   (d) $^{12}_{10}$Mg
   (e) $^{22}_{12}$Mg

7. How many neutrons are in the nucleus of one atom of oxygen–18?
   (a) 2
   (b) 8
   (c) 10
   (d) 18
   (e) 26

Section 5.5 *Atomic Mass*

8. Which of the following isotopes is deflected more than carbon–12 by a magnetic field in a mass spectrometer?
   (a) He–4
   (b) N–14
   (c) O–16
   (d) all of the above
   (e) none of the above

9. Element W occurs naturally as W–6 (6.015 amu) and W–7 (7.016 amu). Calculate the atomic mass of W given that the natural abundance of W–7 is 92.5%.
   (a) 6.09 amu
   (b) 6.50 amu
   (c) 6.52 amu
   (d) 6.94 amu
   (e) 12.5 amu

10. Refer to the periodic table and determine the atomic mass of barium.
    (a) 25 amu
    (b) 56 amu
    (c) 81 amu
    (d) 137.33 amu
    (e) 193.33 amu

11. Refer to the periodic table and determine the mass number for an important isotope of radon.
    (a) 86
    (b) 136
    (c) 222
    (d) 308
    (e) none of the above

Section 5.6 *The Wave Nature of Light*

12. Which of the following colors of light is least energetic?
    (a) blue
    (b) red
    (c) violet
    (d) yellow
    (e) all colors have the same energy

13. Which of the following colors of light has the shortest wavelength?
    (a) blue
    (b) red
    (c) violet
    (d) yellow
    (e) all colors have the same wavelength

14. Which of the following wavelengths of light is least energetic?
    (a) 440 nm
    (b) 470 nm
    (c) 540 nm
    (d) 650 nm
    (e) all wavelengths have the same energy

15. Which of the following wavelengths of light has the lowest frequency?
    (a) 440 nm
    (b) 470 nm
    (c) 540 nm
    (d) 650 nm
    (e) all wavelengths have the same frequency

16. Which of the following wavelengths of light has the fastest velocity?
    (a) 440 nm
    (b) 470 nm
    (c) 540 nm
    (d) 650 nm
    (e) all wavelengths have the same velocity

Section 5.7 *The Quantum Concept*

17. Which of the following instruments gives a quantized measurement?
    (a) 100-mm metric ruler
    (b) 100-g platform balance
    (c) 100-cc hypodermic syringe
    (d) 100-mL volumetric pipet
    (e) none of the above

18. Which of the following is evidence for a quantized change?
    (a) a line spectrum
    (b) a continuous spectrum
    (c) a visible spectrum
    (d) a rainbow
    (e) all of the above

Section 5.8 *Bohr Model of the Atom*

19. Which of these statements is *not* true of the Bohr model of the atom?
    (a) Electrons maintain a constant energy state as they orbit the nucleus.
    (b) Electrons that receive heat energy jump to a higher orbit.
    (c) Electrons gain energy if they drop to an orbit closer to the nucleus.
    (d) Electrons circle the nucleus similar to the way planets circle the sun.
    (e) none of the above

20. Which of the following produces the "atomic fingerprint" for hydrogen?
    (a) atoms of helium reacting
    (b) atoms of helium colliding
    (c) electrons jumping to a higher energy level
    (d) electrons dropping to a lower energy level
    (e) electrons being captured by the nucleus

21. How many photons of light are emitted when the electron in a hydrogen atom drops from energy level 5 to 2?
    (a) 1
    (b) 2
    (c) 3
    (d) 5
    (e) none of the above

Section 5.9 *Principal Energy Levels and Sublevels*

22. How many energy sublevels theoretically exist in the 5th principal energy level?
    - (a) 2
    - (b) 5
    - (c) 10
    - (d) 18
    - (e) none of the above

23. What is the maximum number of electrons that can theoretically occupy the $5d$ sublevel?
    - (a) 2
    - (b) 5
    - (c) 6
    - (d) 10
    - (e) none of the above

24. What is the maximum number of electrons that can occupy the 2nd energy level?
    - (a) 2
    - (b) 6
    - (c) 8
    - (d) 10
    - (e) 18

Section 5.10 *Electron Configuration*

25. What element has the following electron configuration: $1s^2\ 2s^2\ 2p^6\ 3s^2\ 3p^1$?
    - (a) B
    - (b) Na
    - (c) Al
    - (d) K
    - (e) Sc

26. What element has the following electron configuration: $1s^2\ 2s^2\ 2p^6\ 3s^2\ 3p^3$?
    - (a) N
    - (b) Al
    - (c) P
    - (d) K
    - (e) Sc

Section 5.11 *Quantum Mechanical Model of the Atom*

27. Which of the following is a distinction between an orbit and an orbital?
    (a) An orbital represents an energy level.
    (b) An orbital represents a probability boundary.
    (c) An orbital may contain more than one electron.
    (d) An orbital may not gain or lose energy.
    (e) An orbit and an orbital are identical.

28. Which of the following orbitals has a dumbbell shape and is in the ~~fifth~~ 5th shell?
    (a) $5s$
    (b) $4p$
    (c) $5p$
    (d) $4d$
    (e) $5d$

29. What is the maximum number of electrons that can occupy a $4d$ orbital?
    (a) 2
    (b) 3
    (c) 4
    (d) 6
    (e) 10

30. How many orbitals exist within the $4d$ subshell?
    (a) 2
    (b) 4
    (c) 5
    (d) 6
    (e) 10

31. What is the maximum number of electrons that can occupy a $4d$ subshell?
    (a) 2
    (b) 4
    (c) 7
    (d) 10
    (e) 14

32. What is the maximum number of electrons that can occupy the 4th shell?
    (a) 2
    (b) 4
    (c) 10
    (d) 18
    (e) none of the above

Chapter 5 Key Terms

Across

1. the distance light travels in one cycle
3. a spectrum with broad bands of radiant energy
7. an electron orbit designated 1, 2, 3 ... (2 words)
8. the arrangement of electrons in an atom
10. a spectrum with narrow bands of radiant energy
14. a region of high density in the center of the atom
16. an atomic model with electron probability (2 words)
20. the _____ number refers to the A value
23. a formula for a wavelength of light from a H atom
24. a model of an atom with electron orbits
25. a neutral subatomic particle
26. the number of wave cycles in one second

Down

2. the light spectrum from 400–700 nm
4. instrument for determining the mass of an isotope
5. a region of high probability for finding an electron
6. expresses the composition of a nucleus (2 words)
9. 1/12 the mass of a carbon-12 atom
11. atoms that differ only by the number of neutrons
12. the _____ number refers to the Z value
13. a spectrum from X rays to microwaves (2 words)
15. a discrete energy level in an atom
17. the principle that the precise location and energy of an electron cannot both be determined
18. a negatively charged subatomic particle
19. the weighted average mass of isotopes (2 words)
21. an energy level that is designated $s, p, d, f ...$
22. a positively charged subatomic particle

Chapter 5 Answers to Self–Test

1. **b** 2. **a** 3. **d** 4. **b** 5. **c** 6. **b** 7. **c** 8. **a** 9. **d** 10. **d** 11. **c** 12. **b** 13. **c** 14. **d** 15. **d** 16. **e** 17. **d** 18. **a** 19. **c** 20. **d** 21. **a** 22. **b** 23. **d** 24. **d** 25. **c** 26. **c** 27. **b** 28. **c** 29. **a** 30. **c** 31. **d** 32. **e**

Chapter 5 Answers to Crossword Puzzle

Across
1. wavelength
3. continuous
7. energy level
8. configuration
10. line
14. nucleus
16. quantum mechanical
20. mass
23. Balmer
24. Bohr
25. neutron
26. frequency

Down
2. visible
4. spectrometer
5. orbital
6. atomic notation
9. amu
11. isotopes
12. atomic
13. radiant energy
15. quantum
17. uncertainty
18. electron
19. atomic mass
21. sublevel
22. proton

The Periodic Table

Section 6.1 *Classification of Elements*

1. Which of the following scientists did not contribute to the early development of the periodic table of elements?
   - (a) J.W. Döbereiner
   - (b) Dmitri Mendeleev
   - (c) H.G.J. Moseley
   - (d) J.A.R. Newlands
   - (e) Ernest Rutherford

2. Which of the following scientists proposed the law of octaves to explain the repeating properties of the elements?
   - (a) Johann Döbereiner
   - (b) Dmitri Mendeleev
   - (c) Lothar Meyer
   - (d) H.G.J. Moseley
   - (e) J.A.R. Newlands

Section 6.2 *The Periodic Law Concept*

3. The original periodic law was based upon which of the following?
   - (a) increasing atomic mass
   - (b) increasing atomic number
   - (c) increasing isotopic mass
   - (d) increasing mass number
   - (e) increasing neutron number

4. The modern periodic law is based on which of the following?
   - (a) increasing atomic mass
   - (b) increasing atomic number
   - (c) increasing isotopic mass
   - (d) increasing mass number
   - (e) increasing neutron number

Section 6.3 *Groups and Periods of Elements*

5. Which fourth period representative element has the lowest atomic number?
   (a) C
   (b) K
   (c) Sc
   (d) Ga
   (e) Y

6. Which fourth period transition element has the lowest atomic number?
   (a) C
   (b) K
   (c) Sc
   (d) Ga
   (e) Y

7. Previous to the 1985 recommendation by IUPAC, what was the designation for the nitrogen family of elements according to the American convention?
   (a) IIIA
   (b) IIIB
   (c) V A
   (d) V B
   (e) none of the above

Section 6.4 *Periodic Trends*

8. Which of the following is a general trend from left to right in the periodic table?
   (a) atomic radius increases; metallic character increases
   (b) atomic radius increases; metallic character decreases
   (c) atomic radius decreases; metallic character increases
   (d) atomic radius decreases; metallic character decreases
   (e) none of the above

9. Predict which of the following elements has the largest atomic radius.
   (a) F
   (b) Cl
   (c) O
   (d) S
   (e) Se

10. Predict which of the following elements has the most nonmetallic character.
    (a) F
    (b) Cl
    (c) O
    (d) S
    (e) Se

Section 6.5 *Properties of Elements*

11. Predict the density of barium, given the density of calcium (1.54 g/cm³) and strontium (2.63 g/cm³).
    (a) 0.45 g/cm³
    (b) 1.09 g/cm³
    (c) 2.09 g/cm³
    (d) 3.72 g/cm³
    (e) 4.17 g/cm³

12. Predict the atomic radius of antimony given the radius of arsenic (0.125 nm) and bismuth (0.155 nm).
    (a) 0.030 nm
    (b) 0.095 nm
    (c) 0.140 nm
    (d) 0.185 nm
    (e) 0.280 nm

13. Predict the boiling point for argon given the boiling points of krypton (−152°C) and xenon (−107°C).
    (a) −197°C
    (b) −130°C
    (c) −62°C
    (d) −45°C
    (e) 45°C

14. Predict which of the following elements has chemical properties most similar to iron.
    (a) Al
    (b) Ru
    (c) Na
    (d) Ni
    (e) Zn

Section 6.6 *Blocks of Elements*

15. Which energy sublevel is being filled by the elements Ga through Kr?
    (a) $3d$
    (b) $4s$
    (c) $4p$
    (d) $4d$
    (e) $4f$

16. Which energy sublevel is being filled by the elements Y through Cd?
    (a) $4d$
    (b) $5s$
    (c) $5p$
    (d) $5d$
    (e) $5f$

17. Which of the following is the electron configuration for an atom of tin?
    (a) [Kr] $5s^2 4d^{10} 4p^2$
    (b) [Kr] $5s^2 4d^{10} 5p^2$
    (c) [Kr] $5s^2 4d^{10} 5d^2$
    (d) [Xe] $5s^2 4d^{10} 5p^2$
    (e) [Xe] $5s^2 4d^{10} 5d^2$

Section 6.7 *Valence Electrons*

18. Predict the number of valence electrons for a Group IIA/2 element.
    (a) 1
    (b) 2
    (c) 3
    (d) 6
    (e) 8

19. Predict the number of valence electrons for an atom of selenium.
    (a) 2
    (b) 4
    (c) 6
    (d) 16
    (e) 34

Section 6.8 *Electron Dot Formulas*

20. Which of the following is the electron dot formula for an atom of strontium?
    (a) Sr·   (b) S̤r·   (c) ·S̤r: (d) ·S̤r: (e) :S̤r:

21. Which of the following is the electron dot formula for an atom of oxygen?
    (a) O·   (b) Ọ·   (c) ·Ọ·   (d) ·Ọ: (e) :Ọ:

Section 6.9 *Ionization Energy*

22. Which of the following groups of elements has the highest ionization energy?
    (a) Group IA/1
    (b) Group IIA/2
    (c) Group IIB/12
    (d) Group VIIA/17
    (e) Group VIIIA/18

23. Which of the following elements has the lowest ionization energy?
    (a) Rb
    (b) Sr
    (c) Cs
    (d) Ba
    (e) Xe

24. Which of the following is a general trend for the ionization energy of elements in the periodic table?
    (a) increases from left to right; increases from top to bottom
    (b) increases from left to right; decreases from top to bottom
    (c) decreases from left to right; increases from top to bottom
    (d) decreases from left to right; decreases from top to bottom
    (e) none of the above

Section 6.10 *Ionic Charges*

25. Which of the following groups has a predictable ionic charge of two negative?
    (a) Group IIA/2
    (b) Group IVB/4
    (c) Group IVA/14
    (d) Group VIA/16
    (e) Group VIIIA/17

26. Which of the following ions is not isoelectronic with argon?
    (a) S^{2-}
    (b) Cl^-
    (c) K^+
    (d) Sc^{3+}
    (e) V^{4+}

27. What is the electron configuration for a titanium ion, Ti^{4+}?
    (a) $1s^2\ 2s^2\ 2p^6\ 3s^2\ 3p^6$
    (b) $1s^2\ 2s^2\ 2p^6\ 3s^2\ 3p^6\ 4s^2$
    (c) $1s^2\ 2s^2\ 2p^6\ 3s^2\ 3p^6\ 3d^2$
    (d) $1s^2\ 2s^2\ 2p^6\ 3s^2\ 3p^6\ 4s^2\ 3d^2$
    (e) none of the above

28. What is the electron configuration for an indium ion, In^{3+}?
- (a) $1s^2\ 2s^2\ 2p^6\ 3s^2\ 3p^6\ 4s^2\ 3d^{10}\ 4p^6\ 4d^{10}$
- (b) $1s^2\ 2s^2\ 2p^6\ 3s^2\ 3p^6\ 4s^2\ 3d^{10}\ 4p^6\ 5s^2\ 4d^{10}$
- (c) $1s^2\ 2s^2\ 2p^6\ 3s^2\ 3p^6\ 4s^2\ 3d^{10}\ 4p^6\ 5s^2\ 4d^8$
- (d) $1s^2\ 2s^2\ 2p^6\ 3s^2\ 3p^6\ 4s^2\ 3d^{10}\ 4p^6\ 5s^2\ 4d^{10}\ 5p^1$
- (e) none of the above

29. What is the electron configuration for a sulfide ion, S^{2-}?
- (a) $1s^2\ 2s^2\ 2p^6\ 3s^2\ 3p^2$
- (b) $1s^2\ 2s^2\ 2p^6\ 3s^2\ 3p^4$
- (c) $1s^2\ 2s^2\ 2p^6\ 3s^2\ 3p^6$
- (d) $1s^2\ 2s^2\ 2p^6\ 3s^2\ 3p^6\ 4s^2$
- (e) none of the above

30. What is the electron configuration for a telluride ion, Te^{2-}?
- (a) $1s^2\ 2s^2\ 2p^6\ 3s^2\ 3p^6\ 4s^2\ 3d^{10}\ 4p^6\ 5s^2\ 4d^{10}\ 5p^2$
- (b) $1s^2\ 2s^2\ 2p^6\ 3s^2\ 3p^6\ 4s^2\ 3d^{10}\ 4p^6\ 5s^2\ 4d^{10}\ 5p^4$
- (c) $1s^2\ 2s^2\ 2p^6\ 3s^2\ 3p^6\ 4s^2\ 3d^{10}\ 4p^6\ 5s^2\ 4d^{10}\ 5p^6$
- (d) $1s^2\ 2s^2\ 2p^6\ 3s^2\ 3p^6\ 4s^2\ 3d^{10}\ 4p^6\ 5s^2\ 4d^{10}\ 5p^5\ 6s^1$
- (e) none of the above

Periodic Table Exercise

Match the chemical symbol from the following list to the description of an element in the periodic table.

List of Symbols: As Eu Ti Kr Cl Lr Ce Ca Cs H

Description of Element:
- (a) alkali metal
- (b) alkaline earth metal
- (c) halogen
- (d) noble gas
- (e) lowest atomic mass nonmetal
- (f) lowest atomic mass lanthanide
- (g) highest atomic mass actinide
- (h) rare earth metal
- (i) Group VA/15 semimetal
- (j) Group 4 fourth period element

Chapter 6 Key Terms

Across
2. an element such as F, Cl, Br, I
5. an element such as Sc, Y, La (2 words)
6. properties of elements repeat in a pattern (2 words)
7. charge on an atom after gaining or losing electrons
9. the elements with variable chemical properties
11. Lewis structure (2 words)
14. symbol of a noble gas element
17. the elements filling $4f$ and $5f$ sublevels (2 words)
22. symbol of an element more valuable than gold
23. symbol of the least dense metallic element
24. an atom that has gained or lost electrons
25. symbol of a radioactive Group 1 element
26. a row of elements
27. symbol of an alkaline earth element
28. the nucleus and inner electrons of an atom

Down
1. symbol of a liquid nonmetallic element
3. an element with an atomic number 90-103
4. the unreactive gaseous elements
5. the elements with predictable chemical properties
8. ions with identical electron configurations
10. the metals Mg, Ca, Sr (2 words)
12. the elements following atomic number 92
13. symbol of the most dense metallic element
15. an element with an atomic number 58-71
16. a family of elements
18. the metals Li, Na, K
19. the process of an atom losing an electron
20. the outer s and p electrons
21. a simplified notation for electron configuration

Chapter 6 Answers to Self–Test

1. **a** 2. **e** 3. **a** 4. **b** 5. **b** 6. **c** 7. **c** 8. **d** 9. **e** 10. **a** 11. **d** 12. **c** 13. **a** 14. **b** 15. **c** 16. **a**
17. **b** 18. **b** 19. **c** 20. **b** 21. **d** 22. **e** 23. **c** 24. **b** 25. **d** 26. **e** 27. **a** 28. **a** 29. **c** 30. **c**

Periodic Table Exercise

(a)	alkali metal	Cs
(b)	alkaline earth metal	Ca
(c)	halogen	Cl
(d)	noble gas	Kr
(e)	lowest atomic mass nonmetal	H
(f)	lowest atomic mass lanthanide	Ce
(g)	highest atomic mass actinide	Lr
(h)	rare earth metal	Eu
(i)	Group VA/15 semimetal	As
(j)	Group 4 fourth period element	Ti

Chapter 6 Answers to Crossword Puzzle

Across
2. halogen
5. rare earth
6. periodic law
7. ionic
9. transition
11. electron dot
14. Ar
17. inner transition
22. Pt
23. Li
24. ion
25. Fr
26. period
27. Mg
28. kernel

Down
1. Br
3. actinide
4. noble
5. representative
8. isoelectronic
10. alkaline earth
12. transuranium
13. Os
15. lanthanide
16. group
18. alkali
19. ionization
20. valence
21. core

Language of Chemistry

Section 7.1 *Classification of Compounds*

1. The compound $CaSO_4$ is classified as which of the following?
   (a) binary ionic
   (b) ternary ionic
   (c) binary molecular
   (d) binary acid
   (e) ternary oxyacid

2. The compound H_2O is classified as which of the following?
   (a) binary ionic
   (b) ternary ionic
   (c) binary molecular
   (d) binary acid
   (e) ternary oxyacid

3. Aqueous HBr is classified as which of the following?
   (a) binary ionic
   (b) ternary ionic
   (c) binary molecular
   (d) binary acid
   (e) ternary oxyacid

4. The ammonium ion, NH_4^+, is classified as which of the following?
   (a) monoatomic cation
   (b) monoatomic anion
   (c) polyatomic cation
   (d) polyatomic anion
   (e) none of the above

5. The hydroxide ion, OH^-, is classified as which of the following?
   (a) monoatomic cation
   (b) monoatomic anion
   (c) polyatomic cation
   (d) polyatomic anion
   (e) none of the above

Section 7.2 *Monoatomic Ions*

6. What is the systematic IUPAC name for Hg_2^{2+} according to the Stock system?
   (a) mercury ion
   (b) mercury(I) ion
   (c) mercury(II) ion
   (d) mercuric ion
   (e) mercurous ion

7. What is the systematic IUPAC name for Cu^{2+} according to the Latin system?
   (a) copper ion
   (b) copper(II) ion
   (c) cupric ion
   (d) cuprous ion
   (e) cuprum ion

8. What is the systematic IUPAC name for I^-?
   (a) hypoiodite ion
   (b) iodide ion
   (c) iodite ion
   (d) iodate ion
   (e) periodate ion

9. What is the predicted ionic charge of an element in Group VIA/16?
   (a) 2+
   (b) 6+
   (c) 2–
   (d) 6–
   (e) none of the above

Section 7.3 *Polyatomic Ions*

10. What is the systematic IUPAC name for MnO_4^-?
    (a) manganese ion
    (b) manganate ion
    (c) manganite ion
    (d) permanganate ion
    (e) none of the above

11. What is the formula for the hydrogen carbonate ion?
   (a) $C_2H_3O_2^-$
   (b) $C_2H_3O_2^{2-}$
   (c) HCO_3^-
   (d) HCO_3^{2-}
   (e) none of the above

Section 7.4 *Writing Chemical Formulas*

12. What is the formula for the ionic compound composed of the barium ion, Ba^{2+}, and the phosphide ion, P^{3-}?
   (a) Ba_2P_2
   (b) Ba_2P_3
   (c) Ba_3P_2
   (d) Ba_3P_3
   (e) Ba_6P_6

13. What is the formula for the ionic compound composed of the bismuth ion, Bi^{3+}, and the cyanide ion, CN^-?
   (a) $BiCN$
   (b) $BiCN_3$
   (c) Bi_3CN
   (d) $Bi(CN)_3$
   (e) $Bi_3(CN)_3$

Section 7.5 *Binary Ionic Compounds*

14. What is the ionic charge for the cobalt ion in Co_2S_3?
   (a) 2+
   (b) 2–
   (c) 3+
   (d) 3–
   (e) none of the above

15. What is the systematic IUPAC name for Co_2S_3 according to the Stock system?
   (a) cobalt sulfide
   (b) cobalt(II) sulfide
   (c) cobalt(II) sulfite
   (d) cobalt(III) sulfide
   (e) cobalt(III) sulfite

16. Predict the chemical formula for rubidium fluoride, given the formula of sodium fluoride, NaF.
    (a) RbF
    (b) Rb_2F
    (c) RbF_2
    (d) Rb_2F_3
    (e) Rb_3F_2

Section 7.6 *Ternary Ionic Compounds*

17. What is the ionic charge for the cobalt ion in $Co_3(PO_4)_2$?
    (a) 2+
    (b) 2–
    (c) 3+
    (d) 3–
    (e) none of the above

18. What is the systematic IUPAC name for $Co_3(PO_4)_2$ according to the Latin system?
    (a) cobaltic phosphide
    (b) cobaltous phosphite
    (c) cobaltic phosphite
    (d) cobaltous phosphate
    (e) cobaltic phosphate

19. Predict the chemical formula for aluminum selenide, given the formula of aluminum oxide, Al_2O_3.
    (a) AlSe
    (b) Al_2Se
    (c) $AlSe_3$
    (d) Al_2Se_3
    (e) Al_3Se_2

Section 7.7 *Binary Molecular Compounds*

20. What is the systematic IUPAC name for SF_6?
    (a) sulfur tetrafluoride
    (b) sulfur hexafluoride
    (c) sulfur heptafluoride
    (d) sulfur octafluoride
    (e) sulfur nonafluoride

21. What is the formula for tetraphosphorus trisulfide?
    (a) PS_3
    (b) P_4S
    (c) P_3S_4
    (d) P_4S_3
    (e) P_4S_4

Section 7.8 *Binary Acids*

22. What is the systematic IUPAC name for aqueous HBr?
    (a) hydrogen bromide
    (b) hydrobromous acid
    (c) hydrobromic acid
    (d) bromous acid
    (e) bromic acid

23. What is the formula for hydrosulfuric acid?
    (a) $H_2S(aq)$
    (b) $HSO_3(aq)$
    (c) $HSO_4(aq)$
    (d) $H_2SO_3(aq)$
    (e) $H_2SO_4(aq)$

Section 7.9 *Ternary Oxyacids*

24. What is the systematic IUPAC name for aqueous $HC_2H_3O_2$?
    (a) acetic acid
    (b) carbonic acid
    (c) dicarbonic acid
    (d) hydrogen carbonous acid
    (e) hydrogen carbonic acid

25. What is the formula for sulfurous acid?
    (a) $H_2S(aq)$
    (b) $HSO_3(aq)$
    (c) $HSO_4(aq)$
    (d) $H_2SO_3(aq)$
    (e) $H_2SO_4(aq)$

Chemical Nomenclature Exercises

26. Give an acceptable IUPAC name for each of the following chemical formulas.

    (a) KF

    (b) CuO

    (c) $Zn(NO_3)_2$

    (d) $NiSO_4$

    (e) $(NH_4)_2CrO_4$

27. Write the chemical formula for each of the following compounds.

    (a) silver iodide

    (b) cadmium nitride

    (c) manganese(II) chloride

    (d) tin(IV) carbonate

    (e) iron(II) hydroxide

Chapter 7 Key Terms

Across

2. any positively charged ion
4. classification of aqueous H_2SO_4 (2 words)
6. the simplest particle in an ionic compound (2 words)
7. classification of aqueous HCl (2 words)
9. a compound of two nonmetals (2 words)
12. a multiple-atom ion
14. -ous and -ic suffix naming system
16. releases hydroxide ions in water
17. releases hydrogen ions in water
18. Roman numeral naming system

Down

1. a substance not containing carbon
3. a partially neutralized acidic compound (2 words)
4. a compound of a metal and two nonmetals (2 words)
5. any negatively charged ion
7. a compound of a metal and one nonmetal (2 words)
8. an ionic compound from an acid-base reaction
10. a solution of a substance dissolved in water
11. a single-atom ion
13. the simplest particle in a molecular compound
15. International Union of Pure and Applied Chemistry

Chapter 7 Answers to Self–Test

1. **b** 2. **c** 3. **d** 4. **c** 5. **d** 6. **b** 7. **c** 8. **b** 9. **c** 10. **d** 11. **c** 12. **c** 13. **d** 14. **c** 15. **d** 16. **a**
17. **a** 18. **d** 19. **d** 20. **b** 21. **d** 22. **c** 23. **a** 24. **a** 25. **d** 26. **b** 27. **a**

Chemical Nomenclature Exercises

28.   (a)   potassium fluoride   (b)   copper(II) oxide, or cupric oxide
      (c)   zinc nitrate   (d)   nickel(II) sulfate
      (e)   ammonium chromate

29.   (a) AgI (b) Cd_3N_2
      (c) $MnCl_2$ (d) $Sn(CO_3)_2$
      (e) $Fe(OH)_2$

> **Note**: To assist you in learning the names and the formulas of ions, sheets of flashcards are available at the back of this *Study Guide and Solutions Manual*.

Chapter 7 Answers to Crossword Puzzle

Across
  2. cation
  4. ternary oxyacid
  6. formula unit
  7. binary acid
  9. binary molecular
12. polyatomic
14. Latin
16. base
17. acid
18. Stock

Down
  1. inorganic
  3. acid salt
  4. ternary ionic
  5. anion
  7. binary ionic
  8. salt
10. aqueous
11. monoatomic
13. molecule
15. IUPAC

Chemical Reactions

Section 8.1 *Evidence for Chemical Reactions*

1. Which of the following is evidence for a chemical reaction?
   - (a) a gas is detected
   - (b) a precipitate is formed
   - (c) a color change is observed
   - (d) an energy change is noted
   - (e) all of the above

2. Which of the following is not evidence for a chemical reaction producing a gas?
   - (a) an odor is detected
   - (b) a color change is observed
   - (c) a flaming splint is extinguished
   - (d) a glowing splint bursts into flames
   - (e) bubbles are observed

Section 8.2 *Writing Chemical Equations*

3. Which of the statements below best describes the following reaction?

$$2\,AgClO_3(s) \xrightarrow{\Delta} 2\,AgCl(s) + 3\,O_2(g)$$

   - (a) Silver chlorate gives silver chloride and oxygen.
   - (b) Silver chlorate is heated to give silver chloride and oxygen.
   - (c) Silver chlorate gives solid silver chloride and oxygen gas.
   - (d) Silver chlorate is heated to give solid silver chloride and oxygen gas.
   - (e) Solid silver chlorate is heated to give solid silver chloride and oxygen gas.

Section 8.3 *Balancing Chemical Equations*

4. Which of the following is not a general direction for balancing an equation?
   (a) write correct formulas for reactants and products
   (b) begin balancing with the most complex formula
   (c) balance polyatomic ions as a single unit
   (d) balance ionic compounds as a single unit
   (e) check each reactant and product to verify the coefficients

5. What is the coefficient of carbon dioxide after balancing the following equation?

$$\_CuHCO_3(s) \xrightarrow{\Delta} \_Cu_2CO_3(s) + \_H_2O(g) + \_CO_2(g)$$

   (a) 1   (b) 2   (c) 3   (d) 4   (e) 6

6. What is the coefficient of water after balancing the following equation?

$$\_HNO_3(aq) + \_Co(OH)_2(aq) \rightarrow \_Co(NO_3)_2(aq) + \_H_2O(l)$$

   (a) 1   (b) 2   (c) 3   (d) 4   (e) 6

7. What is the coefficient of gold(III) sulfate after balancing the following equation?

$$\_ZnSO_4(aq) + \_Au(NO_3)_3(aq) \rightarrow \_Au_2(SO_4)_3(s) + \_Zn(NO_3)_2(aq)$$

   (a) 1   (b) 2   (c) 3   (d) 4   (e) 6

8. What is the coefficient of lead metal after balancing the following equation?

$$\_Al(s) + \_Pb(C_2H_3O_2)_2(aq) \rightarrow \_Al(C_2H_3O_2)_3(aq) + \_Pb(s)$$

   (a) 1   (b) 2   (c) 3   (d) 4   (e) 6

Section 8.4 *Classifying Chemical Reactions*

9. Classify the type of chemical reaction illustrated in Exercise 8.
   (a) combination
   (b) decomposition
   (c) single replacement
   (d) double replacement
   (e) neutralization

Section 8.5 *Combination Reactions*

10. What is the product predicted from the following combination reaction?

$$Li(s) \ + \ O_2(g) \ \xrightarrow{\Delta}$$

(a) LiO
(b) Li_2O
(c) LiO_2
(d) Li_2O_3
(e) Li_3O_2

11. What is the product predicted from the following combination reaction?

$$S(s) \ + \ O_2(g) \ \xrightarrow{\Delta}$$

(a) SO
(b) S_2O
(c) SO_2
(d) SO_3
(e) It is impossible to predict a single product for a nonmetal and oxygen without more information.

12. What is the product predicted from the following combination reaction?

$$Ni(s) \ + \ Br_2(g) \ \xrightarrow{\Delta}$$

(a) NiBr
(b) Ni_2Br
(c) $NiBr_2$
(d) Ni_2Br_3
(e) Ni_3Br_2

Section 8.6 *Decomposition Reactions*

13. What are the products from the following decomposition reaction?

$$Bi(HCO_3)_3(s) \ \xrightarrow{\Delta}$$

(a) Bi, H_2O, and CO_2
(b) Bi_2CO_3, H_2, and CO_2
(c) $Bi_2(CO_3)_3$ and H_2O
(d) Bi_2CO_3, H_2O, and CO_2
(e) $Bi_2(CO_3)_3$, H_2O, and CO_2

14. What are the products from the following decomposition reaction?

$$Co_2(CO_3)_3(s) \xrightarrow{\Delta}$$

(a) Co and CO_2
(b) CoO and CO_2
(c) Co_2O_3 and CO_2
(d) CoO, H_2O, and CO_2
(e) Co_2O_3, H_2O, and CO_2

15. What are the products from the following decomposition reaction?

$$Ca(ClO_3)_2(s) \xrightarrow{\Delta}$$

(a) Ca and CO_2
(b) Ca, Cl_2, and O_2
(c) $CaCl_2$ and H_2O
(d) $CaCl_2$ and O_2
(e) $CaCl_2$ and CO_2

Section 8.7 *The Activity Series Concept*

16. Which of the following metals reacts with aqueous $FeSO_4$?
(Refer to the Activity Series in Appendix D of the textbook.)
(a) Al
(b) Mg
(c) Zn
(d) all of the above
(e) none of the above

17. Which of the following metals reacts with aqueous $AgNO_3$?
(Refer to the Activity Series in Appendix D of the textbook.)
(a) Hg
(b) Cu
(c) Au
(d) all of the above
(e) none of the above

18. Which of the following metals does not react with sulfuric acid?
(Refer to the Activity Series in Appendix D of the textbook.)
(a) Ag
(b) Cd
(c) Co
(d) Mn
(e) Sn

19. Which of the following metals reacts with water at 25°C?
    (a) Al
    (b) Fe
    (c) Zn
    (d) Mg
    (e) Sr

Section 8.8 *Single–Replacement Reactions*

20. What are the products from the following single–replacement reaction?

$$Mg(s) \quad + \quad AgNO_3(aq) \quad \rightarrow$$

    (a) Ag and $Mg(NO_3)_2$
    (b) Ag and $Mg(NO_2)_2$
    (c) Ag_2O and $Mg(NO_3)_2$
    (d) Ag_2O and $Mg(NO_2)_2$
    (e) no reaction

21. What are the products from the following single–replacement reaction?

$$Zn(s) \quad + \quad H_2SO_4(aq) \quad \rightarrow$$

    (a) ZnO and H_2SO_3
    (b) ZnO and H_2S
    (c) $ZnSO_4$ and H_2
    (d) $ZnSO_4$ and H_2O
    (e) no reaction

22. What are the products from the following single–replacement reaction?

$$Na(s) \quad + \quad H_2O(l) \quad \rightarrow$$

    (a) Na_2O and H_2
    (b) Na_2O and H_2O
    (c) NaOH and H_2
    (d) NaOH and H_2O
    (e) no reaction

Section 8.9 *Solubility Rules*

23. Which of the following solid compounds is insoluble in water?
    (Refer to the Solubility Rules in Appendix E of the textbook.)
    (a) $(NH_4)_2CO_3$
    (b) K_2CrO_4
    (c) $PbSO_4$
    (d) Na_2S
    (e) $Sr(OH)_2$

Section 8.10 *Double–Replacement Reactions*

24. What are the products from the following double replacement reaction?

$$BaCl_2(aq) \ + \ K_2SO_4(aq) \quad \rightarrow$$

   (a) BaS and $KClO_4$
   (b) $BaSO_3$ and KCl
   (c) $BaSO_3$ and $KClO_4$
   (d) $BaSO_4$ and KCl
   (e) $BaSO_4$ and $KClO_4$

Section 8.11 *Neutralization Reactions*

25. What are the products from the following neutralization reaction?

$$HNO_3(aq) \ + \ NH_4OH(aq) \quad \rightarrow$$

   (a) NH_3NO_3 and H_2O
   (b) NH_3NO_3 and O_2
   (c) NH_4NO_3 and H_2O
   (d) NH_4NO_3 and O_2
   (e) NH_4NO_2 and H_2O

Translating Chemical Reactions into Balanced Equations

26. Write a balanced chemical equation for the reaction of aqueous strontium nitrate with aqueous sodium sulfate to yield a precipitate of strontium sulfate and aqueous sodium nitrate.

27. Write a balanced chemical equation for the reaction of hydrochloric acid with aqueous barium hydroxide to give aqueous barium chloride and water.

Chapter 8 Key Terms

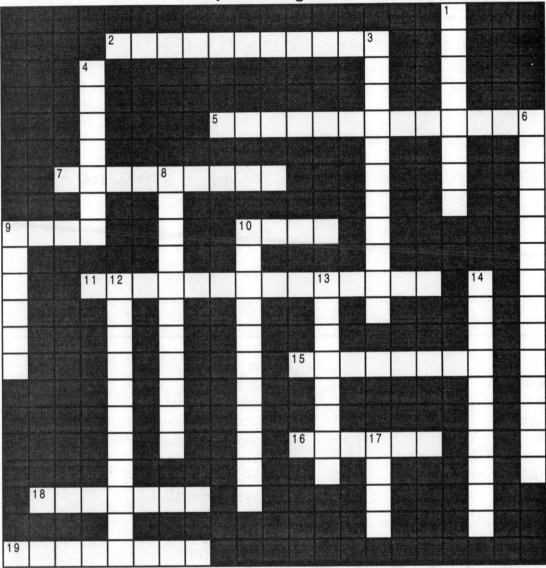

Across
2. an insoluble salt in solution
5. a type of reaction involving a single reactant
7. a number in a chemical formula
9. a compound produced from an acid-base reaction
10. releases hydrogen ions in solution
11. lists the relative ability to react (2 words)
15. increases rate of reaction
16. a type of replacement reaction that switches ions
18. a solution with a substance dissolved in water
19. a reaction requires this substance

Down
1. a chemical change
3. a reaction that absorbs heat
4. a reaction yields this substance
6. a type of reaction involving an acid and base
8. a type of reaction involving two elements
9. a type of replacement reaction of a metal in an acid
10. an element that reacts with water (2 words)
12. a number preceding a formula in an equation
13. a description of a reaction using formulas and symbols
14. a reaction that releases heat
17. releases hydroxide ions in solution

Chapter 8 Answers to Self–Test

1. e 2. b 3. e 4. d 5. a 6. b 7. a 8. c 9. c 10. b 11. e 12. c 13. e 14. c 15. d 16. d
17. b 18. a 19. e 20. a 21. c 22. c 23. c 24. d 25. c

Translating Chemical Reactions into Balanced Equations

26. $Sr(NO_3)_2(aq) + Na_2SO_4(aq) \rightarrow SrSO_4(s) + 2\,NaNO_3(aq)$

27. $2\,HCl(aq) + Ba(OH)_2(aq) \rightarrow BaCl_2(aq) + 2\,H_2O(l)$

Chapter 8 Answers to Crossword Puzzle

Across
2. precipitate
5. decomposition
7. subscript
9. salt
10. acid
11. activity series
15. catalyst
16. double
18. aqueous
19. reactant

Down
1. reaction
3. endothermic
4. product
6. neutralization
8. combination
9. single
10. active metal
12. coefficient
13. equation
14. exothermic
17. base

CHAPTER

The Mole Concept

9

Section 9.1 *Avogadro's Number*

1. How many atoms of carbon have a mass equal to its atomic mass, 12.01 g?
   (a) 1
   (b) 6
   (c) 12
   (d) 6.02×10^{23}
   (e) none of the above

2. What is the mass of Avogadro's number of iron atoms?
   (a) 55.85 amu
   (b) 55.85 grams
   (c) 6.02×10^{23} amu
   (d) 6.02×10^{23} grams
   (e) none of the above

Section 9.2 *Mole Calculations I*

3. Which of the following is equal to one mole of substance?
   (a) 6.02×10^{23} sodium atoms, Na
   (b) 6.02×10^{23} chlorine molecules, Cl_2
   (c) 6.02×10^{23} sodium chloride formula units, NaCl
   (d) 6.02×10^{23} chloride ions, Cl–
   (e) all of the above

4. How many xenon atoms are in 0.250 mol of the noble gas?
   (a) 1.51×10^{22} atoms
   (b) 1.51×10^{23} atoms
   (c) 1.51×10^{24} atoms
   (d) 2.41×10^{23} atoms
   (e) 2.41×10^{24} atoms

5. How many moles of argon correspond to 7.52×10^{22} atoms of the inert gas?
    (a) 0.0125 mol
    (b) 0.0801 mol
    (c) 0.125 mol
    (d) 0.801 mol
    (e) 1.25 mol

Section 9.3 *Molar Mass*

6. What is the approximate molar mass of trinitrotoluene (TNT), $C_7H_5(NO_2)_3$?
    (a) 63.07 g/mol
    (b) 135.13 g/mol
    (c) 189.21 g/mol
    (d) 227.15 g/mol
    (e) 405.39 g/mol

Section 9.4 *Mole Calculations II*

7. What is the mass of 4.50×10^{22} atoms of gold?
    (a) 0.0679 g
    (b) 0.0748 g
    (c) 13.3 g
    (d) 14.7 g
    (e) 197 g

8. How many methane molecules, CH_4, have a mass of 3.20 g?
    (a) 1.20×10^{23} molecules
    (b) 1.20×10^{24} molecules
    (c) 1.93×10^{24} molecules
    (d) 3.01×10^{23} molecules
    (e) 3.01×10^{24} molecules

9. What is the mass of a formula unit of cuprous sulfide, Cu_2S?
    (a) 1.04×10^{-26} g
    (b) 1.66×10^{-24} g
    (c) 2.64×10^{-22} g
    (d) 3.79×10^{21} g
    (e) 9.57×10^{25} g

Section 9.5 *Molar Volume*

10. One mole of which of the following gases occupies 22.4 L at STP?
    (a) ammonia, NH_3
    (b) helium, He
    (c) hydrogen, H_2
    (d) oxygen, O_2
    (e) all of the above

11. What is the density of nitrogen gas, N_2, at STP?
    (a) 0.625 g/L
    (b) 0.800 g/L
    (c) 1.25 g/L
    (d) 1.60 g/L
    (e) 22.4 g/L

12. If the density of nitrous oxide is 1.96 g/L at STP, what is the molar mass of laughing gas?
    (a) 11.4 g/mol
    (b) 19.6 g/mol
    (c) 22.4 g/mol
    (d) 43.9 g/mol
    (e) 196 g/mol

Section 9.6 *Mole Calculations III*

13. What is the volume of 1.51×10^{24} molecules of ammonia gas, NH_3, at STP?
    (a) 0.112 L
    (b) 0.399 L
    (c) 2.51 L
    (d) 8.93 L
    (e) 56.2 L

14. What is the mass of 0.500 liter of oxygen gas, O_2, at STP?
    (a) 0.286 g
    (b) 0.714 g
    (c) 3.50 g
    (d) 6.40 g
    (e) 112 g

15. The formula for the illegal drug cocaine is $C_{17}H_{21}NO_4$. What is the percentage of nitrogen in the compound?
    (a) 4.62%
    (b) 6.99%
    (c) 21.1%
    (d) 32.6%
    (e) 67.3%

16. Phenacyl chloride is a used as tear gas for riot control and has the formula C_7H_7OCl. What is the percentage of carbon in the compound?
    (a) 4.96%
    (b) 8.42%
    (c) 11.2%
    (d) 24.9%
    (e) 59.0%

Section 9.8 *Empirical Formula*

17. If 0.250 mol V reacts with 0.375 mol O, what is the empirical formula of vanadium oxide?
    (a) VO
    (b) V_2O_3
    (c) V_2O_5
    (d) V_3O_2
    (e) V_5O_2

18. If 1.500 g of vanadium metal react with oxygen gas to give 2.679 g of vanadium oxide, what is the empirical formula of the product?
    (a) VO
    (b) V_2O_3
    (c) V_2O_5
    (d) V_3O_2
    (e) V_5O_2

19. Acetaldehyde is used in the manufacture of perfumes, flavors, and dyes. Calculate the empirical formula for acetaldehyde given that its percent composition is: 54.53% C, 9.15% H, and 36.32% O.
    (a) $C_1H_1O_1$
    (b) $C_2H_4O_1$
    (c) $C_2H_3O_1$
    (d) $C_5H_9O_2$
    (e) $C_6H_9O_3$

Section 9.9 *Molecular Formula*

20. Find the molecular formula for aspirin given that its empirical formula is $C_9H_8O_4$ and its approximate molar mass is 175 g/mol.
    (a) $C_1H_1O_1$
    (b) $C_2H_2O_1$
    (c) $C_9H_8O_4$
    (d) $C_{18}H_{16}O_8$
    (e) $C_{27}H_{24}O_{12}$

21. Lindane is an insecticide that causes dizziness and diarrhea in humans. Find the molecular formula for lindane given that its percent composition is: 24.8% C, 2.1% H, and 73.1% Cl. The approximate molar mass of lindane is 290 g/mol.
    (a) $C_1H_1Cl_1$
    (b) $C_3H_3Cl_3$
    (c) $C_1H_1Cl_6$
    (d) $C_6H_6Cl_6$
    (e) $C_{12}H_1Cl_{35}$

Analogies for Avogadro's Number

22. Given 6.02×10^{23} hydrogen atoms (0.07–nm in diameter), placed side-by-side, which of the following is the best estimate of the total length?
    (a) a diameter of a pin
    (b) a centimeter
    (c) a meter stick
    (d) a kilometer
    (e) a million trips around Earth's equator

23. Given 6.02×10^{23} steel shotput balls, which of the following is the best estimate of their total mass?
    (a) a microgram
    (b) a kilogram
    (c) an automobile
    (d) the Earth
    (e) the universe

24. Given 6.02×10^{23} softballs, which of the following is the best estimate of their total volume?
    (a) 1000 mL beaker
    (b) the Los Angeles Coliseum
    (c) the Grand Canyon
    (d) the Earth
    (e) the universe

25. One mole of germanium has a volume of 13.6 cm^3. Using X-ray diffraction, each germanium atom is found to have a cubic volume 0.284 nm on a side. Calculate the experimental value for Avogadro's number (N).
    (a) 1.69×10^{23}
    (b) 2.09×10^{23}
    (c) 4.79×10^{23}
    (d) 5.94×10^{23}
    (e) 6.02×10^{23}

Empirical and Molecular Formulas   Show all work for each of the following.

26. Galactose is the sugar that gives milk a sweet taste and its percent composition is 40.00% C, 6.72% H, 53.29% O. What is the empirical formula of galactose?

27. If the approximate molar mass of galactose is about 180 g/mol, what is the molecular formula of galactose?

Chapter 9 Key Terms

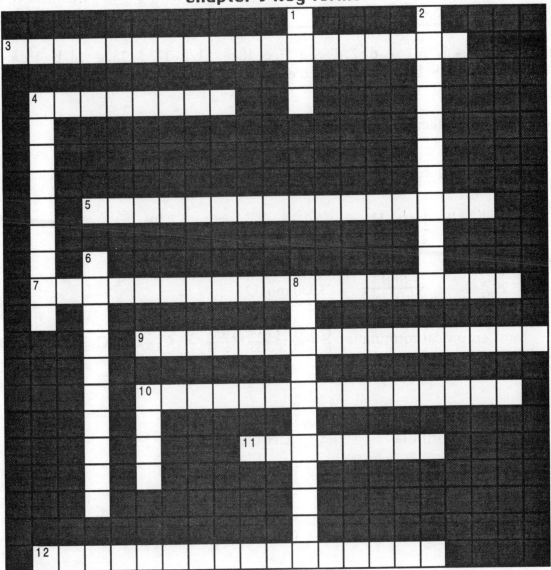

Across

3. a list of elements in a compound by mass (2 words)
4. particle in a molecular compound (2 words)
5. the simplest ratio of atoms in a compound (2 words)
7. 0°C (2 words)
9. the actual ratio of atoms in a compound (2 words)
10. number of atoms in 12 g of carbon-12 (2 words)
11. stated equal volumes of gas have equal molecules
12. one atmosphere (2 words)

Down

1. amount of substance with 6.02×10^{23} particles
2. particle in an ionic compound (2 words)
4. the mass of one mole of substance (2 words)
6. usually expressed in grams per liter (2 words)
8. the space occupied by one mole of gas (2 words)
10. representative particle of an element

Chapter 9 Answers to Self–Test

1. d 2. b 3. e 4. b 5. c 6. d 7. d 8. a 9. c 10. e 11. c 12. d 13. e 14. b 15. a
16. e 17. b 18. c 19. b 20. c 21. d 22. e 23. d 24. d 25. d

Empirical and Molecular Formulas

26. $40.00 \text{ gC} \times \dfrac{1 \text{ mol C}}{12.01 \text{ gC}} = 3.330 \text{ mol C}$

$6.72 \text{ gH} \times \dfrac{1 \text{ mol H}}{1.01 \text{ gH}} = 6.65 \text{ mol H}$

$53.29 \text{ gO} \times \dfrac{1 \text{ mol O}}{16.00 \text{ gO}} = 3.330 \text{ mol O}$

thus, empirical formula is $C_{\frac{3.33}{3.33}} H_{\frac{6.65}{3.33}} O_{\frac{3.33}{3.33}} = C_{1.00}H_{1.99}O_{1.00} = CH_2O$

27. $\dfrac{\text{molecular formula}}{\text{empirical formula}} = \dfrac{(CH_2O)_n}{CH_2O} = n$

$\dfrac{180 \text{ g}}{30 \text{ g}} = 6$

thus, molecular formula is $(CH_2O)_6 = C_6H_{12}O_6$

Chapter 9 Answers to Crossword Puzzle

Across
3. percent composition
4. molecule
5. empirical formula
7. standard temperature
9. molecular formula
10. Avogadro's number
11. Avogadro
12. standard pressure

Down
1. mole
2. formula unit
4. molar mass
6. gas density
8. molar volume
10. atom

Stoichiometry

Section 10.1 *Interpreting a Chemical Equation*

1. How many moles of oxygen gas, O_2, react with 2 moles of nitrogen monoxide gas, NO, according to the following equation?

$$\_NO(g) \ + \ \_O_2(g) \ \xrightarrow{\Delta/UV} \ \_NO_2(g)$$

   (a) 1 mol
   (b) 2 mol
   (c) 3 mol
   (d) 4 mol
   (e) none of the above

2. How many molar masses of nitrogen dioxide gas are produced from 2 molar masses of nitrogen monoxide gas according to the following equation?

$$\_NO(g) \ + \ \_O_2(g) \ \xrightarrow{\Delta/UV} \ \_NO_2(g)$$

   (a) 1 MM
   (b) 2 MM
   (c) 3 MM
   (d) 4 MM
   (e) none of the above

3. How many liters of oxygen gas react to produce 2 L of nitrogen dioxide gas (at similar conditions) according to the following equation?

$$\_NO(g) \ + \ \_O_2(g) \ \xrightarrow{\Delta/UV} \ \_NO_2(g)$$

   (a) 1 L
   (b) 2 L
   (c) 3 L
   (d) 4 L
   (e) none of the above

4. If 0.300 g of nitrogen monoxide produces 0.460 g of nitrogen dioxide, predict the mass of reacting oxygen gas from the conservation of mass law.

$$\_NO(g) \ + \ \_O_2(g) \ \xrightarrow{\Delta/UV} \ \_NO_2(g)$$

(a) 0.0800 g
(b) 0.160 g
(c) 0.320 g
(d) 0.380 g
(e) 0.760 g

Section 10.2 *Mole–Mole Problems*

5. How many moles of water react with 0.500 mol of lithium metal according to the following reaction?

$$\_Li(s) \ + \ \_H_2O(l) \ \rightarrow \ \_LiOH(aq) \ + \ \_H_2(g)$$

(a) 0.125 mol
(b) 0.250 mol
(c) 0.500 mol
(d) 1.00 mol
(e) 2.00 mol

6. How many moles of hydrogen gas are produced from the reaction of 0.500 mol of sodium metal?

$$\_Na(s) \ + \ \_H_2O(l) \ \rightarrow \ \_NaOH(aq) \ + \ \_H_2(g)$$

(a) 0.125 mol
(b) 0.250 mol
(c) 0.500 mol
(d) 1.00 mol
(e) 2.00 mol

7. How many moles of water must react in order to produce 0.500 mol of potassium hydroxide?

$$\_K(s) \ + \ \_H_2O(l) \ \rightarrow \ \_KOH(aq) \ + \ \_H_2(g)$$

(a) 0.125 mol
(b) 0.250 mol
(c) 0.500 mol
(d) 1.00 mol
(e) 2.00 mol

Section 10.3 *Types of Stoichiometry Problems*

8. Classify the following type of stoichiometry problem: *How many grams of sulfur must react with an excess volume of oxygen in order to yield 0.500 L of sulfur dioxide gas?*
   - (a) mass–mass problem
   - (b) mass–volume problem
   - (c) volume–volume problem
   - (d) mole–mole
   - (e) none of the above

9. Classify the following type of stoichiometry problem: *How many grams of sulfur must react with an excess volume of oxygen in order to yield 0.500 g of sulfur dioxide gas?*
   - (a) mass–mass problem
   - (b) mass–volume problem
   - (c) volume–volume problem
   - (d) mole–mole
   - (e) none of the above

10. Classify the following type of stoichiometry problem: *How many liters of sulfur must react with an excess volume of oxygen in order to yield 50.0 cm^3 of sulfur dioxide gas?*
    - (a) mass–mass problem
    - (b) mass–volume problem
    - (c) volume–volume problem
    - (d) mole–mole
    - (e) none of the above

11. In general, how many unit factors are necessary to solve mass–mass stoichiometry problems?
    - (a) one
    - (b) two
    - (c) three
    - (d) four
    - (e) five

12. In general, how many unit factors are necessary to solve mass–volume stoichiometry problems?
    - (a) one
    - (b) two
    - (c) three
    - (d) four
    - (e) five

13. Which of the following steps is *not* necessary to solve a volume–volume stoichiometry problem?
    (a) write a balanced chemical equation for the reaction
    (b) calculate the moles of known gas given its volume
    (c) convert the volume of known gas to the volume of unknown gas
    (d) determine the volume of known gas
    (e) determine the volume of unknown gas

Section 10.4 *Mass–Mass Problems*

14. How many grams of iron are produced from the reaction of 500.0 g of aluminum metal?

$$\_FeO(l) \ + \ \_Al(l) \ \xrightarrow{\Delta} \ \_Fe(l) \ + \ \_Al_2O_3(l)$$

    (a) 345 g
    (b) 689 g
    (c) 1030 g
    (d) 1550 g
    (e) 3100 g

15. How many grams of aluminum metal must react to give 500.0 g of iron?

$$\_FeO(l) \ + \ \_Al(l) \ \xrightarrow{\Delta} \ \_Fe(l) \ + \ \_Al_2O_3(l)$$

    (a) 80.6 g
    (b) 161 g
    (c) 242 g
    (d) 362 g
    (e) 483 g

Section 10.5 *Mass–Volume Problems*

16. What STP volume of oxygen is released from the decomposition of 2.166 g of mercuric oxide (216.59 g/mol)?

$$\_HgO(s) \ \xrightarrow{\Delta} \ \_Hg(l) \ + \ \_O_2(g)$$

    (a) 0.112 L
    (b) 0.224 L
    (c) 0.448 L
    (d) 1.12 L
    (e) 2.24 L

17. How many grams of mercuric oxide (216.59 g/mol) must decompose to release 0.375 L of oxygen gas at STP?

$$\_HgO(s) \xrightarrow{\Delta} \_Hg(l) + \_O_2(g)$$

(a) 0.0752 g
(b) 1.82 g
(c) 3.63 g
(d) 7.25 g
(e) 14.5 g

Section 10.6 *Volume–Volume Problems*

18. Calculate the volume of oxygen gas that reacts with 32.0 mL of sulfur dioxide. (Assume constant conditions.)

$$\_SO_2(g) + \_O_2(g) \xrightarrow{Pt/650°C} \_SO_3(g)$$

(a) 8.00 mL
(b) 16.0 mL
(c) 32.0 mL
(d) 64.0 mL
(e) 96.0 mL

19. Calculate the volume of sulfur dioxide that produces 25.0 mL of sulfur trioxide. (Assume constant conditions.)

$$\_SO_2(g) + \_O_2(g) \xrightarrow{Pt/650°C} \_SO_3(g)$$

(a) 6.25 mL
(b) 12.5 mL
(c) 25.0 mL
(d) 50.0 mL
(e) 100.0 mL

Section 10.7 *The Limiting Reactant Concept*

20. If 1.00 mol of chromium reacts with 1.00 mol of oxygen gas according to the following equation, how many moles of chromium(III) oxide are produced?

$$Cr(s) + O_2(g) \rightarrow Cr_2O_3(s)$$

(a) 0.500 mol
(b) 0.667 mol
(c) 1.00 mol
(d) 1.50 mol
(e) 2.00 mol

Section 10.8 *Limiting Reactant Problems*

21. If 50.0 mL of sulfur dioxide gas reacts with 30.0 mL of oxygen gas, what is the volume of sulfur trioxide gas produced (assume constant conditions)?

$$SO_2(g) + O_2(g) \rightarrow SO_3(g)$$

(a) 10.0 mL
(b) 20.0 mL
(c) 50.0 mL
(d) 60.0 mL
(e) 80.0 mL

22. If 10.0 g of aluminum metal react with 10.0 g of sulfur according to the following equation, how many grams of aluminum sulfide are produced?

$$Al(s) + S(s) \rightarrow Al_2S_3(s)$$

(a) 15.6 g
(b) 20.0 g
(c) 27.8 g
(d) 46.8 g
(e) 55.7 g

Section 10.9 *Percent Yield*

23. Starting with 0.657 g of lead(II) nitrate, a student collects 0.905 g of precipitate. If the calculated mass of precipitate is 0.914 g, what is the percent yield?

(a) 71.9%
(b) 72.6%
(c) 99.0%
(d) 101%
(e) 138%

Stoichiometry Problems   Show all work for each of the following.

24. What is the mass of solid silver chloride produced from the reaction of excess aqueous silver nitrate and 0.333 g of barium chloride?

25. What is the volume of oxygen gas liberated at STP when 1.555 g of solid silver chlorate is decomposed with heat to give solid silver chloride?

Chapter 10 Key Terms

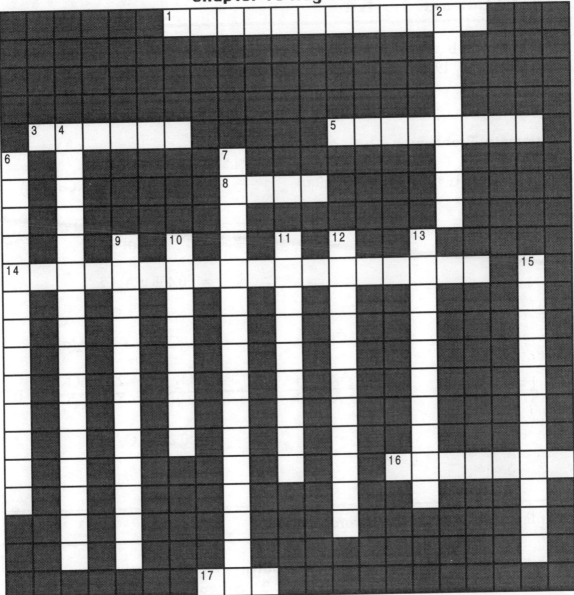

Across
1. liters-to-liters type of problem (2 words)
3. the experimental yield from a reaction
5. grams-to-grams type of problem (2 words)
8. the element produced from the blast furnace process
14. Laviosier's law (3 words)
16. the substance produced from the Haber process
17. 0°C and 1 atm

Down
2. a single particle composed of nonmetal atoms
4. Gay-Lussac's law (2 words)
6. relates quantities according to a balanced equation
7. controls the amount of product (2 words)
9. ratio of actual to theoretical yield × 100 (2 words)
10. stated equal volumes—same number of molecules
11. the mass of one mole of substance (2 words)
12. the volume of one mole of gas (2 words)
13. grams-to-liters type of problem (2 words)
15. the calculated yield from a reaction

Chapter 10 Answers to Self–Test

1. a 2. b 3. a 4. b 5. c 6. b 7. c 8. b 9. a 10. c 11. c 12. c 13. b 14. d 15. b
16. a 17. d 18. b 19. c 20. a 21. c 22. a 23. c

Stoichiometry Problems

24.
$$BaCl_2(s) \ + \ 2\,AgNO_3(aq) \ \rightarrow \ 2\,AgCl(s) \ + \ Ba(NO_3)_2(aq)$$

$$0.333 \ \text{g BaCl}_2 \times \frac{1 \ \text{mol BaCl}_2}{208.23 \ \text{g BaCl}_2} \times \frac{2 \ \text{mol AgCl}}{1 \ \text{mol BaCl}_2} \times \frac{143.32 \ \text{g AgCl}}{1 \ \text{mol AgCl}} = \text{g AgCl}$$

$$= \ 0.458 \ \text{g AgCl}$$

25.
$$2\,AgClO_3(s) \ \xrightarrow{\Delta} \ 2\,AgCl(s) \ + \ 3\,O_2(g)$$

$$1.555 \ \text{g AgClO}_3 \times \frac{1 \ \text{mol AgClO}_3}{191.32 \ \text{g AgClO}_3} \times \frac{3 \ \text{mol O}_2}{2 \ \text{mol AgClO}_3} \times \frac{22.4 \ \text{L O}_2}{1 \ \text{mol O}_2} = \text{L O}_2$$

$$= \ 0.273 \ \text{L O}_2$$

Chapter 10 Answers to Crossword Puzzle

Across
1. volume–volume
3. actual
5. mass–mass
8. iron
14. conservation of mass
16. ammonia
17. STP

Down
2. molecule
4. combining volumes
6. stoichiometry
7. limiting reactant
9. percent yield
10. Avogadro
11. molar mass
12. molar volume
13. mass–volume
15. theoretical

The Gaseous State

Section 11.1 *Properties of Gases*

1. Which of the following is an observed property of gases?
   - (a) gases have a variable shape and volume
   - (b) gases expand infinitely
   - (c) gases have low density
   - (d) gases diffuse and mix
   - (e) all of the above

Section 11.2 *Atmospheric Pressure*

2. Which of the following does not express standard atmospheric pressure?
   - (a) 29.9 in. Hg
   - (b) 14.7 psi
   - (c) 76 cm Hg
   - (d) 760 torr
   - (e) 101 Pa

3. If the pressure of nitrogen gas is 15 mm Hg, what is the pressure in atm?
   - (a) 0.020 atm
   - (b) 0.20 atm
   - (c) 15 atm
   - (d) 1100 atm
   - (e) 11,000 atm

4. If the atmospheric pressure is 30.4 in. Hg, what is the pressure in psi?
   - (a) 14.9 psi
   - (b) 30.4 psi
   - (c) 149 psi
   - (d) 3990 psi
   - (e) 39,900 psi

Section 11.3 *Variables Affecting Gas Pressure*

5. The pressure exerted by a gas is affected by three factors. Which of the following *increases* the pressure?
   (a) increasing the volume
   (b) decreasing the temperature
   (c) increasing the number of gas molecules
   (d) all of the above
   (e) none of the above

Section 11.4 *Boyle's Law*

6. Which of the following graphs represents the plot of pressure versus volume at constant temperature?

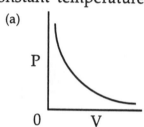
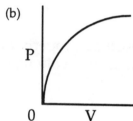
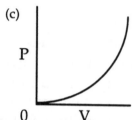
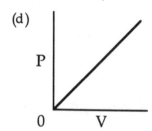
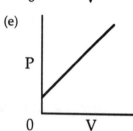

7. A sample of ammonia gas at 1.20 atm compresses from 500 mL to 250 mL. If the temperature remains constant, what is the new pressure of the gas?
   (a) 0.600 atm
   (b) 1.00 atm
   (c) 1.20 atm
   (d) 2.40 atm
   (e) 4.80 atm

Section 11.5 *Charles' Law*

8. A sample of helium gas at 45°C expands from 125 mL to 250 mL. If the pressure remains constant, what is the new Celsius temperature of the gas?
   (a) −114°C
   (b) 23°C
   (c) 159°C
   (d) 363°C
   (e) 636°C

9. Which of the following graphs represents the plot of volume versus Kelvin temperature at constant pressure?

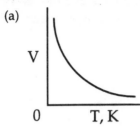 (a)
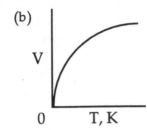 (b)
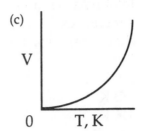 (c)
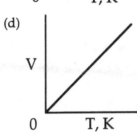 (d)
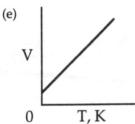 (e)

Section 11.6 *Gay–Lussac's Law*

10. Which of the following graphs represents the plot of pressure versus Kelvin temperature at constant volume?

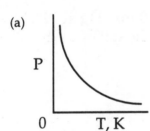 (a)
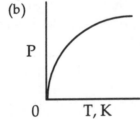 (b)
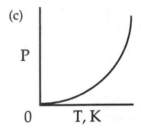 (c)
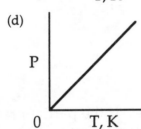 (d)
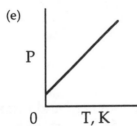 (e)

11. A sample of chlorine gas at 785 mm Hg is heated from 50°C to 100°C. If the volume remains constant, what is the new pressure of the gas?
    (a) 153 mm Hg
    (b) 393 mm Hg
    (c) 680 mm Hg
    (d) 907 mm Hg
    (e) 1570 mm Hg

Section 11.7 *Combined Gas Law*

12. A sample of hydrogen gas is in a cylinder at 0°C. If the gas is heated so that the pressure increases from 1.00 to 2.00 atm, and the volume increases from 2.50 to 5.00L, what is the new temperature of the gas?
    (a) 68 K
    (b) 135 K
    (c) 273 K
    (d) 546 K
    (e) 1092 K

13. A sample of oxygen gas has a volume of 20.0 mL. If the gas cools from 450 K to 225 K, and the pressure decreases from 550 to 275 torr, what is the new volume of the gas?
    (a) 5.00 mL
    (b) 10.0 mL
    (c) 20.0 mL
    (d) 40.0 mL
    (e) 80.0 mL

Section 11.8 *The Vapor Pressure Concept*

14. The vapor pressure of ethanol is 360 mm Hg at 60°C; 600 mm Hg at 72°C; 760 mm Hg at 78°C; 810 mm Hg at 80°C; and 1190 mm Hg at 90°C. What is the normal boiling point of ethanol?
    (a) 60°C
    (b) 72°C
    (c) 78°C
    (d) 80°C
    (e) 90°C

Section 11.9 *Dalton's Law*

15. An air sample contains nitrogen, oxygen, and argon gases. If the partial pressure of nitrogen is 588 torr, oxygen is 158 torr, and argon is 7 torr, what is the pressure of the air sample?
    (a) 7 torr
    (b) 14 torr
    (c) 423 torr
    (d) 753 torr
    (e) 768 torr

16. Oxygen is collected over water at 50.0°C and 770.5 mm Hg. What is the partial pressure of the oxygen gas? (Refer to Appendix F of the textbook for the vapor pressure of water.)
    (a) 678.0 mm Hg
    (b) 720.5 mm Hg
    (c) 820.5 mm Hg
    (d) 863.0 mm Hg
    (e) none of the above

Section 11.10 *Ideal Gas Behavior*

17. Which of the following is not true according to the kinetic theory of gases?
    (a) molecules occupy zero volume
    (b) molecules are attracted to one another
    (c) molecules move randomly
    (d) molecules have elastic collisions
    (e) molecules have a kinetic energy proportional to temperature

18. At the same temperature, which of the following gases has the fastest velocity?
    (a) hydrogen
    (b) helium
    (c) oxygen
    (d) fluorine
    (e) The average velocity for each gas is the same.

19. What is the temperature at which an ideal gas exerts zero pressure?
    (a) 0°C
    (b) 273°C
    (c) 0 K
    (d) 273 K
    (e) –273 K

Section 11.11 *Ideal Gas Law*

20. If 1.00 mol of krypton gas exerts a pressure of 1.00 atm at 100°C, what is the volume of the gas? ($R = 0.0821$ atm·L/mol·K)
    (a) 2.20 mL
    (b) 8.21 L
    (c) 12.2 L
    (d) 30.6 L
    (e) 1220 L

21. If an unknown gas occupies a volume of 1.50 L at 21°C and 0.950 atm and has a mass of 2.01 g, what is the molar mass of the gas? (R = 0.0821 atm•L/mol•K)
    (a) 19.0 g/mol
    (b) 30.7 g/mol
    (c) 34.0 g/mol
    (d) 69.1 g/mol
    (e) 76.6 g/mol

Combined Gas Law Problems

22. Calcium metal reacts with water to evolve hydrogen gas. If 45.5 mL of hydrogen is collected over water at 20°C and 751.5 mm Hg, what is the volume at STP? (Refer to Appendix F of the textbook for the vapor pressure of water.)

	P	V	T
initial			
final			

23. The thermal decomposition of a potassium chlorate sample gives 5.50 L of oxygen gas at 100°C and 2.25 atm pressure. If the volume is 2.77 L at 7.50 atm, what is temperature of the gas?

	P	V	T
initial			
final			

Chapter 11 Key Terms

Across

4. gas law that relates P, V, and T
5. gas that obeys kinetic theory
6. the lowest possible temperature (2 words)
7. $PV =$ _____
8. pressure exerted by a gas in a mixture (2 words)
14. relates P, V, T, and moles of gas (3 words)
17. gas that deviates from kinetic theory
18. molecular collisions that do not lose energy
19. 0°C and 1 atm
20. SI unit of pressure
21. 1 mm Hg of pressure
22. instrument for measuring atmospheric pressure
23. stated the relationship of gas pressure and volume

Down

1. P and V are _____ proportional
2. gas-liquid equilibrium pressure
3. T and V are _____ proportional
4. stated the relationship of volume and temperature
6. the pressure exerted by molecules in air
8. abbreviation for English unit of pressure
9. 76 cm Hg = 1 _____
10. technique for determining the volume of a gas
11. describes the behavior of an ideal gas (2 words)
12. an empty volume with no gas molecules
13. stated the relationship of pressure and temperature
15. stated pressure of gases in a mixture is additive
16. frequency of molecular collisions (2 words)

Chapter 11 Answers to Self–Test

1. **e** 2. **e** 3. **a** 4. **a** 5. **c** 6. **a** 7. **d** 8. **d** 9. **d** 10. **d** 11. **d** 12. **e** 13. **c** 14. **c** 15. **d**
16. **a** 17. **b** 18. **a** 19. **c** 20. **d** 21. **c**

Combined Gas Law Problems

22.

	P	V	T
initial	$751.5 - 17.5 = 734$ mm Hg	45.5 mL	$20°C + 273 = 293$ K
final	760 mm Hg	V_{final}	273 K

$$45.5 \text{ mL} \times \frac{734 \text{ mm Hg}}{760 \text{ mm Hg}} \times \frac{273 \text{ K}}{293 \text{ K}} = 40.9 \text{ mL}$$

23.

	P	V	T
initial	2.25 atm	5.50 L	$100°C + 273 = 373$ K
final	7.50 atm	2.77 L	T_{final}

$$373 \text{ K} \times \frac{7.50 \text{ atm}}{2.25 \text{ atm}} \times \frac{2.77 \text{ L}}{5.50 \text{ L}} = 626 \text{ K } (353°C)$$

Chapter 11 Answers to Crossword Puzzle

Across
  4. combined
  5. ideal
  6. absolute zero
  7. nRT
  8. partial pressure
  14. ideal gas constant
  17. real
  18. elastic
  19. STP
  20. kilopascal
  21. torr
  22. barometer
  23. Boyle

Down
  1. inversely
  2. vapor
  3. directly
  4. Charles
  6. atmospheric
  8. psi
  9. atmosphere
  10. displacement
  11. kinetic theory
  12. vacuum
  13. Gay–Lussac
  15. Dalton
  16. gas pressure

CHAPTER

12

Chemical Bonding

Section 12.1 *The Chemical Bond Concept*

1. What type of chemical bond results from the attraction between a positive metal ion and a negative nonmetal ion?
   - (a) covalent bond
   - (b) ionic bond
   - (c) metallic bond
   - (d) valence bond
   - (e) none of the above

2. Predict which of the following compounds is held together by covalent bonds.
   - (a) CO_2
   - (b) HgO
   - (c) PbO
   - (d) TiO_2
   - (e) all of the above

Section 12.2 *Ionic Bonds*

3. Potassium metal and bromine vapor react to form the ionic compound potassium bromide. Which of the following statements is true?
   - (a) Valence electrons are transferred from bromine to potassium atoms.
   - (b) The potassium atom is smaller in radius than the potassium ion.
   - (c) Potassium and potassium bromide have similar properties.
   - (d) The smallest representative particle is a molecule.
   - (e) none of the above

4. How many electrons are in the valence shell of a bromine atom before and after reaction with potassium metal to give potassium bromide, KBr?
   (a) 1 and 7, respectively
   (b) 1 and 8, respectively
   (c) 7 and 8, respectively
   (d) 7 and 10, respectively
   (e) 9 and 10, respectively

5. Which noble gas is isoelectronic with a potassium ion?
   (a) neon
   (b) argon
   (c) krypton
   (d) xenon
   (e) radon

6. Which of the following ions has the electron configuration: $1s^2\ 2s^2\ 2p^6\ 3s^2\ 3p^6$ $4s^2\ 3d^{10}\ 4p^6$?
   (a) Br^-
   (b) S^{2-}
   (c) Cs^+
   (d) Sc^{2+}
   (e) La^{3+}

7. What is the predicted electron configuration for the titanium ion, Ti^{4+}?
   (a) $1s^2\ 2s^2\ 2p^6\ 3s^2\ 3p^6\ 4s^2\ 3d^6$
   (b) $1s^2\ 2s^2\ 2p^6\ 3s^2\ 3p^6\ 4s^2\ 3d^4$
   (c) $1s^2\ 2s^2\ 2p^6\ 3s^2\ 3p^6\ 4s^2\ 3d^2$
   (d) $1s^2\ 2s^2\ 2p^6\ 3s^2\ 3p^6\ 3d^2$
   (e) $1s^2\ 2s^2\ 2p^6\ 3s^2\ 3p^6$

Section 12.3 *Covalent Bonds*

8. Powdered yellow phosphorus and reddish-brown bromine vapor react to form red phosphorus bromide. Which of the following statements is not true?
   (a) Valence electrons are shared between phosphorus and iodine atoms.
   (b) There are two bonding electrons distributed over both atoms.
   (c) The bond length is less than the sum of the two atomic radii.
   (d) Phosphorus and phosphorus bromide have similar properties.
   (e) The smallest representative particle is a molecule.

9. How many electrons are in the valence shell of a hydrogen atom before and after reaction with chlorine gas to give hydrogen chloride, HCl?
   (a) 1 and 0, respectively
   (b) 1 and 2, respectively
   (c) 1 and 8, respectively
   (d) 2 and 8, respectively
   (e) 2 and 18, respectively

10. How many electrons are in the valence shell of a chlorine atom before and after the reaction with hydrogen gas to give hydrogen chloride, HCl?
   (a) 1 and 0, respectively
   (b) 1 and 2, respectively
   (c) 1 and 8, respectively
   (d) 7 and 8, respectively
   (e) 7 and 18, respectively

Section 12.4 *Electron Dot Formulas of Molecules*

11. Draw the electron dot formula for silane, SiH_4. How many pairs of nonbonding electrons are there in one molecule of SiH_4?
   (a) 0
   (b) 1
   (c) 2
   (d) 4
   (e) 8

12. Draw the electron dot formula for acetylene, C_2H_4, and state the type of bonds in one molecule.
   (a) 5 single bonds
   (b) 4 single bonds and 1 triple bond
   (c) 4 single bonds and 1 double bond
   (d) 3 single bonds and 1 double bond
   (e) 3 single bonds and 1 triple bond

Section 12.5 *Electron Dot Formulas of Polyatomic Ions*

13. Draw the electron dot formula for the carbonate ion, CO_3^{2-}. How many pairs of nonbonding electrons are there in one carbonate ion?
   (a) 3
   (b) 4
   (c) 8
   (d) 12
   (e) none of the above

14. What type of bonds are in a carbonate ion, CO_3^{2-}?
    (a) 1 single bond and 2 double bonds
    (b) 1 single bond, 1 double bond, and 1 triple bond
    (c) 2 single bonds and 1 double bond
    (d) 2 single bonds and 1 triple bond
    (e) 3 single bonds

Section 12.6 *Polar Covalent Bonds*

15. Which of the following is a general trend for the electronegativity of elements in the periodic table?
    (a) increases from left to right; increases from top to bottom
    (b) increases from left to right; decreases from top to bottom
    (c) decreases from left to right; increases from top to bottom
    (d) decreases from left to right; decreases from top to bottom
    (e) none of the above

16. Which of the following molecules contain polar covalent bonds?
    (a) water, H_2O
    (b) ammonia, NH_3
    (c) methane, CH_4
    (d) laughing gas, N_2O
    (e) all of the above

17. If the electronegativity values for nitrogen and oxygen are 3.0 and 3.5, respectively, what is the bond polarity in a nitric oxide, NO, molecule?
    (a) −0.5
    (b) 0.5
    (c) 3.5
    (d) 6.5
    (e) −6.5

18. Which of the following illustrates the bond polarity in a molecule of nitrogen monoxide, NO?
    (a) $(\delta+)$ N—O $(\delta+)$
    (b) $(\delta+)$ N—O $(\delta-)$
    (c) $(\delta-)$ N—O $(\delta+)$
    (d) $(\delta-)$ N—O $(\delta-)$
    (e) (δ) N—O (δ)

Section 12.7 *Nonpolar Covalent Bonds*

19. Which of the following is an example of a nonpolar covalent molecule?
    - (a) water, H_2O
    - (b) ammonia, NH_3
    - (c) phosphine, PH_3
    - (d) hydrogen chloride, HCl
    - (e) all of the above

20. Which of the following elements does not occur naturally as a nonpolar diatomic molecule?
    - (a) N_2
    - (b) O_2
    - (c) F_2
    - (d) Ne_2
    - (e) H_2

Section 12.8 *Coordinate Covalent Bonds*

21. The NO molecule is converted to NO_2 by adding an oxygen atom to the unshared pair of electrons on the nitrogen atom. What is the term for the resulting bond between nitrogen and oxygen?
    - (a) ionic
    - (b) nonpolar covalent
    - (c) polar covalent
    - (d) coordinate covalent
    - (e) none of the above

22. Which of the following polyatomic ions does not contain a coordinate covalent bond?
    - (a) Hg_2^{2+}
    - (b) H_3O^+
    - (c) NH_4^+
    - (d) ClO_2^-
    - (e) none of the above

Section 12.9 *Shapes of Molecules*

23. What is the molecular shape of a hydrogen sulfide, H_2S, molecule?
    - (a) angular
    - (b) linear
    - (c) tetrahedral
    - (d) trigonal
    - (e) none of the above

24. What is the electron pair geometry for a phosphine, PH_3, molecule?
    (a) angular
    (b) linear
    (c) tetrahedral
    (d) trigonal
    (e) none of the above

25. What is the molecular shape of a phosphine, PH_3, molecule?
    (a) angular
    (b) linear
    (c) tetrahedral
    (d) trigonal
    (e) none of the above

Electron Dot Formulas for Molecules and Polyatomic Ions

26. Calculate the total number of valence electrons for each of the following molecules and draw the electron dot and structural formulas.

    (a) **PI_3**

    (b) **HOCN**

27. Calculate the total number of valence electrons for each of the following polyatomic ions and draw the electron dot and structural formulas.

    (a) **$SO_3{}^{2-}$**

    (b) **$BO_3{}^{3-}$**

Chapter 12 Key Terms

Across
1. the molecular shape of H_2O
5. valence electrons that are shared
10. the 8 valence electrons principle (2 words)
12. an ammonium ion or sulfate ion
13. the number of kilocalories to break a bond
14. a bond containing 4 electrons
18. the ability to attract shared electrons
22. a covalent bond with unequal sharing
23. the electrons in s and p sublevels
24. a bond resulting from the attraction of ions
25. a theory that explains molecular shape
26. a bond resulting from the sharing of electrons
27. a formula that represents bonds with dashes

Down
2. formed by two atoms bonded to a central atom
3. a bond containing 2 electrons
4. the interaction of electrons from two atoms
6. a covalent bond with equal sharing
7. valence electrons that are not shared
8. a formula that portrays valence electrons (2 words)
9. a particle held together by covalent bonds
11. a particle held together by ionic bonds (2 words)
15. a bond containing 6 electrons
16. a sodium ion or chloride ion
17. the notation that indicates bond polarity
19. covalent bond distance between two nuclei
20. a covalent bond with donated electrons
21. the molecular shape of CH_4

Chapter 12 Answers to Self–Test

1. b 2. a 3. e 4. c 5. b 6. a 7. e 8. d 9. b 10. d 11. a 12. c 13. c 14. c 15. b 16. e
17. b 18. b 19. c 20. d 21. d 22. a 23. a 24. c 25. d

Electron Dot Formulas for Molecules and Polyatomic Ions

26.

	Molecule	Valence Electrons	Electron Dot	Structural
(a)	PI_3	$5 + 3(7) = 26 \ e^-$	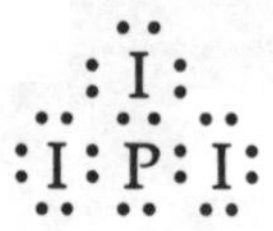	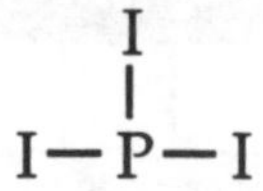
(b)	HOCN	$1 + 6 + 4 + 5 = 16 \ e^-$	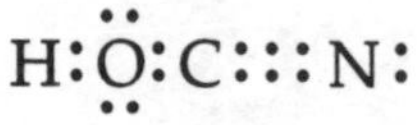	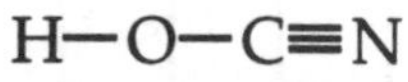

27.

	Ion	Valence Electrons	Electron Dot	Structural
(a)	SO_3^{2-}	$4(6) + 2 = 26 \ e^-$	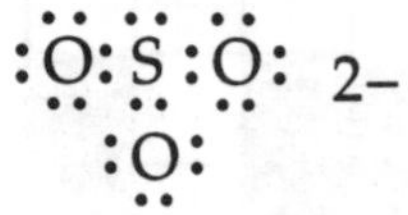	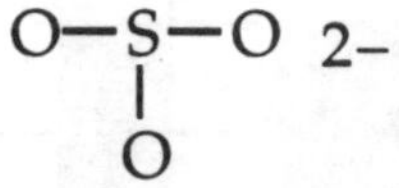
(b)	BO_3^{3-}	$3 + 3(6) + 3 = 24 \ e^-$	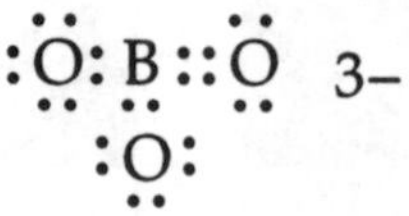	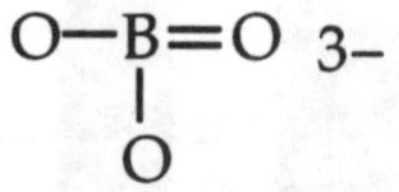

Chapter 12 Answers to Crossword Puzzle

Across
1. angular
5. bonding
10. octet rule
12. polyatomic
13. energy
14. double
18. electronegativity
22. polar
23. valence
24. ionic
25. VSEPR
26. covalent
27. structural

Down
2. angle
3. single
4. bond
6. nonpolar
7. nonbonding
8. electron dot
9. molecule
11. formula unit
15. triple
16. monoatomic
17. delta
19. length
20. coordinate
21. tetrahedral

Liquids and Solids

Section 13.1 *Properties of Liquids*

1. Which of the following properties is not a general characteristic of liquids?
   - (a) definite shape and variable volume
   - (b) flow readily
   - (c) do not compress and expand significantly
   - (d) about 1000 times more dense than gases
   - (e) soluble liquids mix homogeneously

2. Predict the physical state of argon at –190°C (Mp = –189°C, Bp = –186°C) and normal atmospheric pressure.
   - (a) gas
   - (b) liquid
   - (c) solid
   - (d) solid and liquid
   - (e) liquid and gas

Section 13.2 *Vapor Pressure, Viscosity, Surface Tension*

3. Consider the following liquids with similar molar masses. Which liquid has the strongest attraction between molecules based only on boiling point data?
   - (a) Liquid X (Bp = 50°C @ 760 mm Hg)
   - (b) Liquid Y (Bp = 0°C @ 760 mm Hg)
   - (c) Liquid Z (Bp = –50°C @ 760 mm Hg)
   - (d) all liquids have the same intermolecular attraction
   - (e) insufficient data to predict

4. Consider the following liquids with similar molar masses. Which liquid has the strongest attraction between molecules based on vapor pressure data?
   (a) Liquid X (vapor pressure = 50 mm Hg @ 20°C)
   (b) Liquid Y (vapor pressure = 100 mm Hg @ 20°C)
   (c) Liquid Z (vapor pressure = 150 mm Hg @ 20°C)
   (d) all liquids have the same intermolecular attraction
   (e) insufficient data to predict

5. Consider the following liquids with similar molar masses. Which liquid has the strongest attraction between molecules based on viscosity data?
   (a) Liquid X (viscosity = 0.50 centipoise @ 20°C)
   (b) Liquid Y (viscosity = 0.75 centipoise @ 20°C)
   (c) Liquid Z (viscosity = 1.25 centipoise @ 20°C)
   (d) all liquids have the same intermolecular attraction
   (e) insufficient data to predict

6. Consider the following liquids with similar molar masses. Which liquid has the strongest attraction between molecules based on surface tension data?
   (a) Liquid X (surface tension = 15 dynes/cm @ 20°C)
   (b) Liquid Y (surface tension = 25 dynes/cm @ 20°C)
   (c) Liquid Z (surface tension = 50 dynes/cm @ 20°C)
   (d) all liquids have the same intermolecular attraction
   (e) insufficient data to predict

Section 13.3 *The Intermolecular Bond Concept*

7. What is the strongest intermolecular force in a liquid containing polar molecules?
   (a) dispersion forces
   (b) induced dipole forces
   (c) permanent dipole forces
   (d) covalent bonds
   (e) temporary dipole forces

8. What is the strongest intermolecular force in a liquid containing nonpolar molecules?
   (a) dipole forces
   (b) dispersion forces
   (c) hydrogen bonds
   (d) covalent bonds
   (e) none of the above

9. What is the strongest intermolecular force in a liquid having molecules with H–N bonds?
   (a) dipole forces
   (b) dispersion forces
   (c) hydrogen bonds
   (d) covalent bonds
   (e) none of the above

10. Given the following liquids have similar molar masses, which has the strongest molecular attraction?
   (a) $CH_3–CO–O–H$
   (b) $CH_3–CH_2–O–CH_3$
   (c) $CH_3–CH_2–S–CH_3$
   (d) $CH_3–CH_2–CH_2–Cl$
   (e) $CH_3–CH_2–CH_2–CH_2–CH_3$

Section 13.4 *Properties of Solids*

11. Which of the following properties is not a general characteristic of solids?
   (a) crystalline and noncrystalline structures
   (b) do not compress or expand significantly
   (c) heterogeneous solids mix by diffusion
   (d) rigid shape and fixed volume
   (e) usually more dense than corresponding liquids

12. Predict the physical state of argon at –180°C (Mp = –189°C, Bp = –186°C) and normal atmospheric pressure.
   (a) gas
   (b) liquid
   (c) solid
   (d) solid and liquid
   (e) liquid and gas

Section 13.5 *Crystalline Solids*

13. Which of the following is an example of an ionic crystalline solid?
   (a) calcite, $CaCO_3$
   (b) fluorite, CaF_2
   (c) iron pyrite, FeS_2
   (d) silver chloride, $AgCl$
   (e) all of the above

14. Which of the following is not an example of a molecular crystalline solid?
    (a) dry ice, CO_2
    (b) rose gold, Au/Cu
    (c) sucrose, $C_{12}H_{22}O_{11}$
    (d) sulfur, S_8
    (e) urea, $CO(NH_2)_2$

15. Which of the following is an example of a metallic crystalline solid?
    (a) diamond, C
    (b) halite, NaCl
    (c) iodine, I_2
    (d) phosphorus, P_4
    (e) solder, Pb/Sn

Section 13.6 *Changes of Physical State*

16. Calculate the heat required when 10.0 g of ice at –10.0°C warms to water at 20°C. The specific heat of water is 1.00 cal/g×°C; the specific heat of ice is 0.50 cal/g×°C; and the heat of fusion is 80.0 cal/g.
    (a) 105 cal
    (b) 825 cal
    (c) 1050 cal
    (d) 1200 cal
    (e) 2800 cal

17. Calculate the heat released when 25.0 g of steam at 100.0°C cools to water at 0°C. The specific heat of water is 1.00 cal/g×°C; and the heat of vaporization is 540.0 cal/g.
    (a) 4500 cal
    (b) 13,500 cal
    (c) 15,500 cal
    (d) 16,000 cal
    (e) 18,000 cal

Section 13.7 *Structure of Water*

18. What is the number of nonbonding electron pairs in a water molecule?
    (a) 0   (b) 1   (c) 2   (d) 3   (e) 4

19. What is the experimentally observed bond angle in a water molecule?
    (a) 90°
    (b) 104.5°
    (c) 109.5°
    (d) 120°
    (e) 180°

20. Which of the following illustrates the bond polarity in a water molecule?
   (a) $(\delta-)$ O—H $(\delta+)$
   (b) $(\delta-)$ O—H $(\delta-)$
   (c) $(\delta+)$ O—H $(\delta+)$
   (d) $(\delta+)$ O—H $(\delta-)$
   (e) (δ) O—H (δ)

Section 13.8 *Physical Properties of Water*

21. Which of the following accounts for the unusually high heats of fusion and vaporization for water?
   (a) the hydrogen bonds in water
   (b) the molar mass of water
   (c) the specific heat of water
   (d) the surface tension in water
   (e) the viscosity of water

22. Which of the following physical properties has a higher value for water (H_2O) than the corresponding value for heavy water, (D_2O)?
   (a) density
   (b) melting point
   (c) heat of vaporization
   (d) molar mass
   (e) none of the above

Section 13.9 *Chemical Properties of Water*

23. What is(are) the product(s) from the reaction of water and a metal oxide?
   (a) a metal hydroxide
   (b) a metal hydroxide and hydrogen peroxide
   (c) a metal hydroxide and hydrogen gas
   (d) a metal hydroxide and oxygen gas
   (e) a metal hydroxide and ozone gas

24. What are the products from the complete combustion of a hydrocarbon?
   (a) carbon and water
   (b) carbon monoxide and hydrogen
   (c) carbon monoxide and water
   (d) carbon dioxide and hydrogen
   (e) carbon dioxide and water

Section 13.10 *Hydrates*

25. What is the systematic name for $Ca(NO_3)_2 \cdot 4\ H_2O$?
    - (a) calcium nitrate dihydrate
    - (b) calcium nitrate tetrahydrate
    - (c) calcium nitrite dihydrate
    - (d) calcium nitrite tetrahydrate
    - (e) none of the above

26. What is the chemical formula for potassium carbonate decahydrate?
    - (a) $KHCO_3 \cdot H_2O$
    - (b) $KHCO_3 \cdot 10\ H_2O$
    - (c) $K_2CO_3 \cdot H_2O$
    - (d) $K_2CO_3 \cdot 10\ H_2O$
    - (e) $KC_2H_3O_2 \cdot 10\ H_2O$

Reactions of Water as a Reactant or Product

27. Complete and balance each of the following chemical reactions.

$$K(s)\ +\ H_2O(l)\ \rightarrow$$

$$CaO(s)\ +\ H_2O(l)\ \rightarrow$$

$$CO_2(g)\ +\ H_2O(l)\ \rightarrow$$

$$HNO_3(aq)\ +\ Zn(OH)_2(s)\ \rightarrow$$

Chemical Formula of a Hydrate

28. An unknown hydrate of chromium(III) acetate, $Cr(C_2H_3O_2)_3 \cdot X\ H_2O$, is heated to give 7.29% water. Calculate the water of crystallization (X) for the salt and state the formula and name of the hydrate.

Chapter 13 Key Terms

Across
4. the heat required to convert a liquid to a gas
8. a crystalline solid composed of ions
9. the resolution of two or more dipoles
10. a crystalline solid composed of molecules
15. the tendency of a liquid to form drops (2 words)
17. a substance containing a water molecules
20. water containing sodium ions and various anions
23. reaction by passing electricity through a solution
24. the heat required to convert a solid to a liquid
25. water containing a variety of cations and anions

Down
1. temperature when vapor pressure is 1 atm
2. attractive force from temporary dipoles
3. pressure when condensation equals evaporation
5. an oxide that gives an acidic solution
6. a crystalline solid composed of metal atoms
7. arrangement of two atoms bonded to a central atom
11. an intermolecular bond between H and O
12. water of crystallization
13. the resistance of a liquid to flow
14. a solid with particles that repeat in a pattern
16. a compound that does not contain water
18. attractive force from permanent dipoles
19. demineralized water
21. an oxide that gives a basic solution
22. a molecule of water containing deuterium

Chapter 13 Answers to Self–Test

1. **a** 2. **c** 3. **a** 4. **a** 5. **c** 6. **c** 7. **c** 8. **b** 9. **c** 10. **a** 11. **c** 12. **a** 13. **e** 14. **b** 15. **e**
16. **c** 17. **d** 18. **c** 19. **b** 20. **a** 21. **a** 22. **e** 23. **a** 24. **e** 25. **b** 26. **d**

Reactions of Water as a Reactant or Product

27.

$$2\,K(s) \;+\; 2\,H_2O(l) \;\rightarrow\; 2\,KOH(aq) \;+\; H_2(g)$$
$$CaO(s) \;+\; H_2O(l) \;\rightarrow\; Ca(OH)_2(aq)$$
$$CO_2(g) \;+\; H_2O(l) \;\rightarrow\; H_2CO_3(aq)$$
$$2\,HNO_3(aq) \;+\; Zn(OH)_2(s) \;\rightarrow\; Zn(NO_3)_2(aq) \;+\; 2\,H_2O(l)$$

Chemical Formula of a Hydrate

28.

$$Cr(C_2H_3O_2)_3 \cdot X\,H_2O(s) \;\xrightarrow{\Delta}\; Cr(C_2H_3O_2)_3(s) \;+\; X\,H_2O(g)$$

$$7.29\ \cancel{g\,H_2O} \;\times\; \frac{1\ mol\ H_2O}{18.02\ \cancel{g\,H_2O}} \;=\; 0.405\ mol\ H_2O$$

$$92.71\ \cancel{g\,Cr(C_2H_3O_2)_3} \;\times\; \frac{1\ mol\ Cr(C_2H_3O_2)_3}{229.15\ \cancel{g\,Cr(C_2H_3O_2)_3}} \;=\; 0.4046\ mol\ Cr(C_2H_3O_2)_3$$

Water of Crystallization: $\dfrac{0.405}{0.4046} \;=\; 1.00 \approx 1$

Formula: $Cr(C_2H_3O_2)_3 \cdot H_2O$ **Name:** chromium(III) acetate monohydrate

Chapter 13 Answers to Crossword Puzzle

Across
4. vaporization
8. ionic
9. net
10. molecular
15. surface tension
17. hydrate
20. soft
23. electrolysis
24. fusion
25. hard

Down
1. Bp
2. dispersion
3. vapor
5. nonmetal
6. metallic
7. angle
11. hydrogen
12. hydration
13. viscosity
14. crystalline
16. anhydrous
18. dipole
19. deionized
21. metal
22. heavy

Solutions

Section 14.1 *Gases in Solution*

1. Coca-Cola® is carbonated by injecting the liquid with carbon dioxide gas. Under what conditions is the solubility of carbon dioxide gas the greatest?
   (a) low temperature, low pressure
   (b) low temperature, high pressure
   (c) low temperature, pressure is not a factor
   (d) high pressure, temperature is not a factor
   (e) high temperature, high pressure

2. If the solubility of nitrogen in blood is 1.90 mL per 100 mL at one atmosphere, what is the solubility of nitrogen gas at a depth of 155 feet where the pressure is 5.50 atmospheres?
   (a) 0.190 mL/100 mL
   (b) 0.345 mL/100 mL
   (c) 1.90 mL/100 mL
   (d) 3.80 mL/100 mL
   (e) 10.5 mL/100 mL

Section 14.2 *Liquids in Solution*

3. Which of the following illustrates the *like dissolves like* rule for two liquids?
   (a) a polar solvent is insoluble in a polar solvent
   (b) a polar solvent is soluble in a nonpolar solvent
   (c) a nonpolar solvent is soluble in a nonpolar solvent
   (d) all of the above
   (e) none of the above

4. Apply the *like dissolves like* rule to predict which of the following liquids is miscible with water.
   (a) chloroform, $CHCl_3$
   (b) glycerin, $C_3H_5(OH)_3$
   (c) toluene, $C_6H_5CH_3$
   (d) all of the above
   (e) none of the above

Section 14.3 *Solids in Solution*

5. Which of the following illustrates the *like dissolves like* rule for a solid solute in a liquid solvent?
   (a) an ionic compound is insoluble in a polar solvent
   (b) a polar compound is insoluble in a nonpolar solvent
   (c) a nonpolar compound is insoluble in a nonpolar solvent
   (d) all of the above
   (e) none of the above

6. Apply the *like dissolves like* rule to predict which of the following vitamins is insoluble in water.
   (a) thymine, vitamin B_1, $C_{12}H_{18}Cl_2N_4OS$
   (b) riboflavin, vitamin B_2, $C_{17}H_{20}N_4O_6$
   (c) ascorbic acid, vitamin C, $C_6H_8O_6$
   (d) calciferol, vitamin D, $C_{27}H_{44}O$
   (e) all of the above

7. Apply the *like dissolves like* rule to predict which of the following is insoluble in benzene, C_6H_6.
   (a) benzopyrene (a compound in charcoal), $C_{20}H_{12}$
   (b) DDT (an insecticide), $C_{14}H_9Cl_5$
   (c) glycine (an amino acid), $CH_2(NH_2)COOH$
   (d) naphthalene (mothballs), $C_{10}H_8$
   (e) paradichlorobenzene (mothballs), $C_6H_4Cl_2$

Section 14.4 *The Dissolving Process*

8. When fructose, $C_6H_{12}O_6$, dissolves in water, which of the following is formed in solution?
   (a) acetic acid, $HC_2H_3O_2$
   (b) carbonic acid, H_2CO_3
   (c) hydrated clusters of CO_2 molecules
   (d) hydrated clusters of $C_6H_{12}O_6$ molecules
   (e) hydrated clusters of hydrogen and oxygen molecules

9. When solid potassium fluoride dissolves in water, which of the following is one of the aqueous ions formed?

(a) $K^+ \text{ --- } H\text{--}O$
            $|$
            H

(b) $K^+ \text{ --- } O\text{--}H$
            $|$
            H

(c) $F^- \text{ --- } O\text{--}H$
            $|$
            H

(d) all of the above

(e) none of the above

Section 14.5 *Rate of Dissolving*

10. Which of the following increases the rate of dissolving for a solid solute in a solvent?
    (a) grinding the solute
    (b) heating the solution
    (c) stirring the solution
    (d) all of the above
    (e) none of the above

Section 14.6 *Solubility and Temperature*

11. What is the solubility of sucrose sugar, $C_{12}H_{22}O_{11}$, at 20°C? *(Refer to textbook Figure 14.5.)*
    (a) 55 g/100 g water
    (b) 75 g/100 g water
    (c) 90 g/100 g water
    (d) 100 g/100 g water
    (e) 130 g/100 g water

12. What is the minimum temperature required to dissolve 140 g of sucrose sugar, $C_{12}H_{22}O_{11}$, in 100 g of water? *(Refer to textbook Figure 14.5.)*
    (a) 20°C
    (b) 50°C
    (c) 55°C
    (d) 60°C
    (e) 100°C

Section 14.7 *Unsaturated, Saturated, Supersaturated*

13. How concentrated is a solution containing 100 g of $NaC_2H_3O_2$ in 100 g water at 75°C? *(Refer to textbook Figure 14.6.)*
    (a) unsaturated
    (b) saturated
    (c) supersaturated
    (d) superunsaturated
    (e) none of the above

14. How concentrated is a solution containing 100 g of $NaC_2H_3O_2$ in 100 g water at 25°C? *(Refer to textbook Figure 14.6.)*
   (a) unsaturated
   (b) saturated
   (c) supersaturated
   (d) superunsaturated
   (e) none of the above

Section 14.8 *Mass Percent Concentration*

15. Which of the following is not a unit factor related to a 5.00% aqueous solution of sodium hydroxide, NaOH?

   (a) $\dfrac{5.00 \text{ g NaOH}}{95.0 \text{ g solution}}$ (b) $\dfrac{95.0 \text{ g water}}{5.00 \text{ g NaOH}}$

   (c) $\dfrac{5.00 \text{ g NaOH}}{95.0 \text{ g water}}$ (d) $\dfrac{100.0 \text{ g solution}}{5.00 \text{ g NaOH}}$

   (e) $\dfrac{95.0 \text{ g water}}{100.0 \text{ g solution}}$

16. What is the mass of a 10.0% blood plasma solution that contains 2.50 g of dissolved solute?
   (a) 0.250 g
   (b) 0.278 g
   (c) 22.5 g
   (d) 25.0 g
   (e) 250 g

17. What is the mass of glucose, $C_6H_{12}O_6$, solute in 10.0 g of a 5.00% sugar solution?
   (a) 0.180 g
   (b) 0.500 g
   (c) 0.900 g
   (d) 9.50 g
   (e) 10.0 g

18. What is the mass of water needed to prepare 5.00 kg of a 40.0% antifreeze solution?
   (a) 2.00 kg
   (b) 3.00 kg
   (c) 3.33 kg
   (d) 12.5 kg
   (e) 200 kg

Section 14.9 *Molar Concentration*

19. Which of the following is **not** a unit factor related to a 0.500 M NaOH solution?

(a) $\dfrac{0.500 \text{ mol NaOH}}{1 \text{ L solution}}$

(b) $\dfrac{1 \text{ L solution}}{0.500 \text{ mol NaOH}}$

(c) $\dfrac{0.500 \text{ mol NaOH}}{1000 \text{ mL solution}}$

(d) $\dfrac{1000 \text{ mL solution}}{0.500 \text{ mol NaOH}}$

(e) $\dfrac{0.500 \text{ mol NaOH}}{1000 \text{ g solution}}$

20. What is the mass of barium hydroxide (171.35 g/mol) dissolved in 250.0 mL of 0.200 M Ba(OH)$_2$ solution?
    (a) 8.57 g
    (b) 17.1 g
    (c) 85.7 g
    (d) 171 g
    (e) 857 g

21. What volume of 6.00 M sulfuric acid contains 9.80 g of H$_2$SO$_4$ (98.03 g/mol)?
    (a) 0.600 mL
    (b) 16.7 mL
    (c) 60.0 mL
    (d) 167 mL
    (e) 1670 mL

Section 14.10 *Molal Concentration*

22. Which of the following is not a unit factor related to a 0.500 m NaOH solution?

(a) $\dfrac{0.500 \text{ mol NaOH}}{1 \text{ L water}}$

(b) $\dfrac{1 \text{ kg water}}{0.500 \text{ mol NaOH}}$

(c) $\dfrac{0.500 \text{ mol NaOH}}{1000 \text{ g water}}$

(d) $\dfrac{1000 \text{ g water}}{0.500 \text{ mol NaOH}}$

(e) $\dfrac{0.500 \text{ mol NaOH}}{1 \text{ kg water}}$

Section 14.11 *Colligative Properties*

23. Calculate the freezing point for an antifreeze solution that contains 242 g of ethylene glycol, $HOCH_2CH_2OH$, dissolved in 1.15 kg of water ($K_f = 1.86°C/m$).
    - (a) −0.548°C
    - (b) −1.82°C
    - (c) −2.41°C
    - (d) −3.39°C
    - (e) −6.31°C

24. A solution contains 25.0 g of a nonionized compound dissolved in 100.0 g of acetic acid ($K_f = 3.90°C/m$). If the freezing point of acetic acid is lowered 10.0°C, what is the molar mass of the compound?
    - (a) 39.0 g/mol
    - (b) 64.1 g/mol
    - (c) 97.7 g/mol
    - (d) 156 g/mol
    - (e) 975 g/mol

Mass Percent and Molar Concentrations

25. A 50.0 mL sample of solution has a mass of 53.810 g and contains 5.379 g of solid potassium iodide. Calculate the mass percent concentration of the KI solution.

26. Given the data in the preceding problem, calculate the molar concentration of the KI solution.

Chapter 14 Key Terms

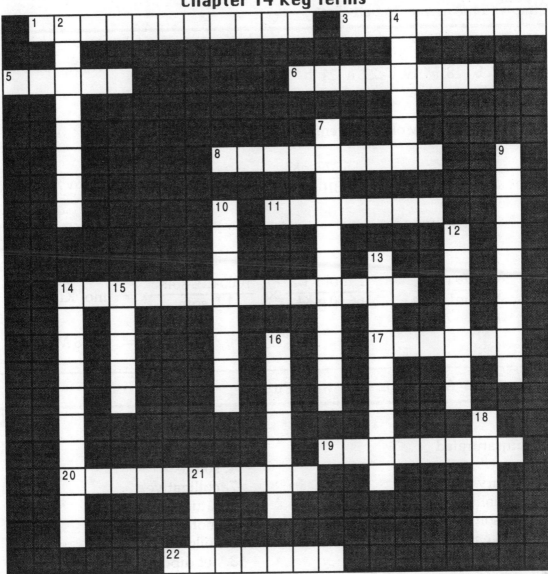

Across

1. a solution with less than maximum solute
3. liquids that are soluble in one another
5. stated gas solubility is proportional to pressure
6. moles of solute dissolved in a kg of solvent
8. like _____ like rule
11. the greater component of a solution
14. a solution with more than maximum solute
17. a molecule with partial charges
19. a solute dissolved in a solvent
20. liquids that separate into two layers
22. grams of solute dissolved in 100 grams of solution

Down

2. a solvent composed of nonpolar molecules
4. the lesser component of a solution
7. a property affected by particles in solution
9. a solution with the maximum solute
10. moles of solute dissolved in each liter of solution
12. a mixture of 1–100 nm dispersed particles
13. the overall direction of negative charge (2 words)
14. the maximum solute in a solvent
15. a solvent composed of polar molecules
16. a type of light scattering by colloid particles
18. _____ freezing point depression constant
21. solvent molecules surrounding a solute particle

Chapter 14 Answers to Self–Test

1. b 2. e 3. c 4. b 5. b 6. d 7. c 8. d 9. b 10. d 11. d 12. c 13. a 14. c 15. a
16. d 17. b 18. b 19. e 20. a 21. b 22. a 23. e 24. c

Mass Percent and Molar Concentrations

25.

$$\frac{5.379 \text{ g KI}}{53.810 \text{ g soln}} \times 100 = \text{m/m \% KI}$$

$$= 9.996\% \text{ KI}$$

26.

$$\frac{5.379 \text{ g KI}}{50.0 \text{ mL soln}} \times \frac{1 \text{ mol KI}}{166.00 \text{ g KI}} \times \frac{1000 \text{ mL soln}}{1 \text{ L soln}} = \text{mol KI/L soln}$$

$$= 0.648 \, M \text{ KI}$$

Chapter 14 Answers to Crossword Puzzle

Across
  1. unsaturated
  3. miscible
  5. Henry
  6. molality
  8. dissolves
  11. solvent
  14. supersaturated
  17. dipole
  19. solution
  20. immiscible
  22. percent

Down
  2. nonpolar
  4. solute
  7. colligative
  9. saturated
  10. molarity
  12. colloid
  13. net dipole
  14. solubility
  15. polar
  16. Tyndall
  18. molal
  21. cage

Acids and Bases

Section 15.1 *Properties of Acids and Bases*

1. Which of the following is a general property of an acidic solution?
   - (a) tastes sour
   - (b) turns blue litmus paper red
   - (c) pH less than 7
   - (d) reacts with a basic solution to give a salt and water
   - (e) all of the above

2. If a solution has a pH of 11, which of the following describes the solution?
   - (a) strongly acidic
   - (b) weakly acidic
   - (c) neutral
   - (d) weakly basic
   - (e) strongly basic

Section 15.2 *Arrhenius Acids and Bases*

3. Given the ionization, which of the following is a strong Arrhenius acid?
   - (a) $HC_2H_3O_2(aq)$ ($\sim$ 1% ionized)
   - (b) $HClO_2(aq)$ ($\sim$ 1% ionized)
   - (c) $H_2SO_3(aq)$ ($\sim$ 1% ionized)
   - (d) $H_3PO_4(aq)$ ($\sim$ 1% ionized)
   - (e) none of the above

4. What acid and base are neutralized to give the salt potassium chloride?
   - (a) $HCl(aq)$ and $KOH(aq)$
   - (b) $HCl(aq)$ and $KCl(aq)$
   - (c) $HOH(aq)$ and $KCl(aq)$
   - (d) $HOH(aq)$ and $KOH(aq)$
   - (e) none of the above

Section 15.3 *Brønsted–Lowry Acids and Bases*

5. In the following reaction, which reactant is acting as a Brønsted–Lowry acid?

$$Na_2HPO_4(aq) \ + \ H_2CO_3(aq) \ \ \rightarrow \ \ NaHCO_3(aq) \ + \ NaH_2PO_4(aq)$$

   (a) Na_2HPO_4
   (b) H_2CO_3
   (c) $NaHCO_3$
   (d) NaH_2PO_4
   (e) H_2O

6. In the following reaction, which reactant is acting as a Brønsted–Lowry base?

$$NaHCO_3(aq) \ + \ NaH_2PO_4(aq) \ \ \rightarrow \ \ Na_2HPO_4(aq) \ + \ H_2CO_3(aq)$$

   (a) $NaHCO_3$
   (b) NaH_2PO_4
   (c) Na_2HPO_4
   (d) H_2CO_3
   (e) H_2O

Section 15.4 *Acid–Base Indicators*

7. Which of the following acid-base indicators is red in an acidic solution and yellow in a basic solution?
   (a) methyl red
   (b) bromthymol blue
   (c) litmus solution
   (d) phenolphthalein
   (e) none of the above

8. Which of the following acid-base indicators is colorless in an acidic solution and pink in a basic solution?
   (a) methyl red
   (b) bromthymol blue
   (c) litmus paper
   (d) phenolphthalein
   (e) none of the above

Section 15.5 *Acid–Base Titrations*

9. If 10.0 mL of 0.500 M $HC_2H_3O_2$ is titrated with 0.250 M NaOH, what volume of sodium hydroxide solution is required to neutralize the acid?

$$HC_2H_3O_2(aq) + NaOH(aq) \rightarrow NaC_2H_3O_2(aq) + H_2O(l)$$

    (a) 5.00 mL
    (b) 10.0 mL
    (c) 20.0 mL
    (d) 40.0 mL
    (e) 80.0 mL

10. If a 0.500 M $HC_2H_3O_2$ solution has a density of 1.00 g/mL, what is the mass percent concentration of the acetic acid solution?
    (a) 2.00%
    (b) 3.00%
    (c) 5.00%
    (d) 10.0%
    (e) 50.0%

Section 15.6 *Acid–Base Standardization*

11. What is the molarity of a potassium hydroxide solution if 25.00 mL of KOH is required to neutralize 2.042 g of $KHC_8H_4O_4$ (204.23 g/mol)?

$$KHC_8H_4O_4(aq) + KOH(aq) \rightarrow K_2C_8H_4O_4(aq) + H_2O(l)$$

    (a) 0.200 M
    (b) 0.250 M
    (c) 0.400 M
    (d) 0.500 M
    (e) 0.800 M

12. Vitamin C has the chemical name ascorbic acid and can be abbreviated HAsc. If 32.00 mL of 0.400 M NaOH neutralizes 2.240 g of ascorbic acid, what is the molar mass of vitamin C?
    (a) 28.0 g/mol
    (b) 87.4 g/mol
    (c) 175 g/mol
    (d) 180 g/mol
    (e) 350 g/mol

Section 15.7 *Ionization of Water*

13. Which of the following is the ionization constant expression for water?
    (a) $K_w = [H_2O][H_2O]$
    (b) $K_w = [H_2O][H_3O^+]/[OH^-]$
    (c) $K_w = [H_2O][OH^-]/[H_3O^+]$
    (d) $K_w = [H_3O^+][OH^-]$
    (e) $K_w = [H_3O^+][OH^-]/[H_2O]$

14. Given an aqueous solution in which the $[H^+] = 5.0 \times 10^{-5}$, what is the molar hydroxide ion concentration?
    (a) $[OH^-] = 2.0 \times 10^{-8}$
    (b) $[OH^-] = 2.0 \times 10^{-9}$
    (c) $[OH^-] = 2.0 \times 10^{-10}$
    (d) $[OH^-] = 5.0 \times 10^{-9}$
    (e) $[OH^-] = 5.0 \times 10^{-10}$

15. Which of the following is true if the hydrogen ion concentration of an aqueous solution increases?
    (a) the pH increases
    (b) the K_w increases
    (c) the K_w decreases
    (d) the hydroxide ion concentration increases
    (e) the hydroxide ion concentration decreases

Section 15.8 *The pH Concept*

16. What is the pH of a solution with a 0.001 molar hydrogen ion concentration?
    (a) 1
    (b) 2
    (c) 3
    (d) 4
    (e) 10

17. What is the molar hydrogen ion concentration of stomach acid that registers a pH of 2 on a strip of pH paper?
    (a) $0.02\ M$
    (b) $0.01\ M$
    (c) $0.2\ M$
    (d) $0.1\ M$
    (e) $2\ M$

Section 15.9 *Advanced pH Calculations*

18. What is the pH of a sample solution if the $[H^+] = 0.000\,065\,M$?
    (a) 0.81
    (b) 4.19
    (c) 4.81
    (d) 5.19
    (e) 7.81

19. What is the $[H^+]$ of a sample solution if the pH is 10.30?
    (a) $1.01\,M$
    (b) $0.000\,000\,000\,20\,M$
    (c) $0.000\,000\,000\,50\,M$
    (d) $0.000\,000\,000\,020\,M$
    (e) $0.000\,000\,000\,050\,M$

Section 15.10 *Strong and Weak Electrolytes*

20. If a light bulb in a conductivity apparatus glows brightly when testing a solution, which of the following is true?
    (a) The solution is highly ionized.
    (b) The solution is slightly ionized.
    (c) The solution is nonionized.
    (d) The solution is highly reactive.
    (e) The solution is slightly reactive.

21. Which of the following aqueous solutions are strong electrolytes?
    (a) strong acids
    (b) strong bases
    (c) soluble salts
    (d) all of the above
    (e) none of the above

22. Sodium chloride is a soluble salt and silver chloride is a slightly soluble salt. Which of the following correctly portrays aqueous solutions of the two salts?
    (a) NaCl(aq) *and* AgCl(s)
    (b) Na$^+$(aq) + Cl$^-$(aq) *and* AgCl(s)
    (c) NaCl(aq) *and* Ag$^+$(aq) + Cl$^-$(aq)
    (d) Na$^+$(aq) + Cl$^-$(aq) *and* Ag$^+$(aq) + Cl$^-$(aq)
    (e) none of the above

Section 15.11 *Net Ionic Equations*

23. Write a balanced net ionic equation given the following total ionic equation:

$$H^+(aq) + NO_3^-(aq) + NH_4OH(aq) \rightarrow NH_4^+(aq) + NO_3^-(aq) + H_2O(l)$$

(a)	$HNO_3(aq) + NH_4OH(aq)$	$\rightarrow$	$NH_4NO_3(aq) + H_2O(l)$
(b)	$H^+(aq) + OH^-(aq)$	$\rightarrow$	$H_2O(l)$
(c)	$NO_3^-(aq) + NH_4^+(aq)$	$\rightarrow$	$NH_4NO_3(aq)$
(d)	$H^+(aq) + NH_4OH(aq)$	$\rightarrow$	$NH_4^+(aq) + H_2O(l)$
(e)	$HNO_3(aq) + OH^-(aq)$	$\rightarrow$	$NO_3^-(aq) + H_2O(l)$

Total and Net Ionic Equations

24. Write the total and net ionic equations for the following reaction.

$$H_2SO_4(aq) + 2\,NaOH(aq) \rightarrow Na_2SO_4(s) + 2\,H_2O(l)$$

Acid–Base Titration

25. A 0.950 g sample of solid sodium carbonate is titrated with hydrochloric acid using bromcresol green indicator. Calculate the molar concentration of the acid if 38.10 mL of HCl are required to completely neutralize the sodium carbonate.

$$HCl(aq) + Na_2CO_3(s) \rightarrow NaCl(aq) + H_2O(l) + CO_2(g)$$

Chapter 15 Key Terms

Across

3. a polar compound dissolving in water
5. a Brønsted–Lowry base is a proton _____
8. ions that do not appear in the net ionic equation
10. a substance that can accept or donate a proton
12. an ionic equation showing all ionized substances
13. an ionic compound dissolving in water
17. a substance that releases hydroxide ions in water
19. an ionic equation with no spectator ions
20. a solution that resists changes in pH
21. an Arrhenius acid releases _____ ions
22. the hydrogen ion in aqueous solution

Down

1. a Brønsted–Lowry acid is a proton _____
2. an electrolyte that is a poor conductor of electricity
3. a substance that changes color with pH
4. the product of H^+ and OH^- concentrations
6. the molar H^+ expressed on an exponential scale
7. a product from a neutralization reaction
9. a solution whose concentration is known precisely
11. an Arrhenius base donates _____ ions
14. an electrolyte that is a good conductor
15. a substance that releases hydrogen ions in water
16. a procedure for delivering a solution with a buret
18. an indicator color change during a titration

Chapter 15 Answers to Self–Test

1. e   2. d   3. e   4. a   5. b   6. a   7. a   8. d   9. c   10. b   11. c   12. c   13. d   14. c   15. e
16. c   17. b   18. b   19. e   20. a   21. d   22. b   23. d

Total Ionic and Net Ionic Equations

24. $\quad\quad\quad\quad H_2SO_4(aq) + 2\,NaOH(aq) \rightarrow Na_2SO_4(aq) + 2\,H_2O(l)$

$2\,H^+(aq) + SO_4{}^{2-}(aq) + 2\,Na^+(aq) + 2\,OH^-(aq) \rightarrow 2\,Na^+(aq) + SO_4{}^{2-}(aq) + 2\,H_2O(l)$

$$H^+(aq) + OH^-(aq) \rightarrow H_2O(l)$$

Acid–Base Titration

25. $\quad\quad\quad\quad 2\,HCl(aq) + Na_2CO_3(s) \rightarrow 2\,NaCl(aq) + H_2O(l) + CO_2(g)$

$$0.950 \; \cancel{g\,Na_2CO_3} \times \frac{1 \; \cancel{mol\,Na_2CO_3}}{105.99 \; \cancel{g\,Na_2CO_3}} \times \frac{2 \; mol\,HCl}{1 \; \cancel{mol\,Na_2CO_3}} = mol\,HCl$$

$$\frac{0.0179 \; mol\,HCl}{38.10 \; \cancel{mL\,soln}} \times \frac{1000 \; \cancel{mL\,soln}}{1 \; L\,soln} = mol\,HCl/L\,soln$$

$$= 0.470 \; M \; HCl$$

Chapter 15 Answers to Crossword Puzzle

Across
- 3. ionization
- 5. acceptor
- 8. spectator
- 10. amphiprotic
- 12. total
- 13. dissociation
- 17. base
- 19. net
- 20. buffer
- 21. hydrogen
- 22. hydronium

Down
- 1. donor
- 2. weak
- 3. indicator
- 4. constant
- 6. pH
- 7. salt
- 9. standard
- 11. hydroxide
- 14. strong
- 15. acid
- 16. titration
- 18. endpoint

Chemical Equilibrium

Section 16.1 *Collision Theory*

1. Which of the following factors affects the rate of a chemical reaction?
   (a) collision frequency
   (b) collision energy
   (c) collision geometry
   (d) all of the above
   (e) none of the above

2. Which of the following increases the collision frequency of molecules?
   (a) increasing the temperature; adding a catalyst
   (b) decreasing the temperature; adding a catalyst
   (c) decreasing the concentration; adding a catalyst
   (d) increasing the concentration; decreasing the temperature
   (e) increasing the concentration; increasing the temperature

3. Which of the following increases the collision energy of molecules?
   (a) increasing the temperature
   (b) increasing the concentration
   (c) increasing the collision frequency
   (d) decreasing the volume of the container
   (e) adding a catalyst

4. Which of the following factors increases the rate of a chemical reaction?
   (a) decreasing the concentration of reactants
   (b) decreasing the temperature
   (c) adding a catalyst
   (d) all of the above
   (e) none of the above

5. Which of the following increases the amount of product from a reaction?
   (a) adding a metal catalyst
   (b) adding an acid catalyst
   (c) using an UV light catalyst
   (d) all of the above
   (e) none of the above

Section 16.2 *Energy Profiles of Chemical Reactions*

6. Which of the is the following is an energy profile for an *exothermic* chemical reaction?

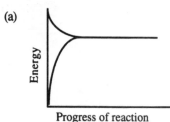
(a)

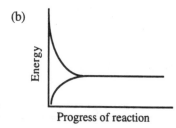
(b)

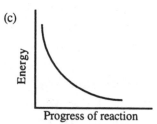
(c)

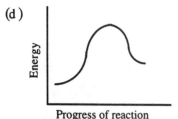
(d)

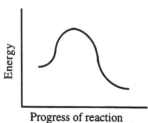
(e)

7. If the heat of reaction is exothermic, which of the following is always true?
   (a) the reaction rate is fast
   (b) the reaction rate is slow
   (c) the energy of the reactants is less than the products
   (d) the energy of the reactants is greater than the products
   (e) none of the above

8. If the E_{act} is lowered, which of the following is always true?
   (a) the reaction is exothermic
   (b) the reaction is endothermic
   (c) the reaction proceeds faster
   (d) the reaction proceeds slower
   (e) none of the above

9. Which of the following lowers the ΔH for a chemical reaction?
   (a) increasing reactant concentration
   (b) increasing product concentration
   (c) adding a catalyst
   (d) all of the above
   (e) none of the above

Section 16.3 *The Chemical Equilibrium Concept*

10. Which of the following is true before a reaction reaches chemical equilibrium?
    (a) The rate of the forward reaction is increasing; the rate of the reverse reaction is decreasing.
    (b) The rate of the forward reaction is decreasing; the rate of the reverse reaction is increasing.
    (c) The rates of the forward and reverse reactions are increasing.
    (d) The rates of the forward and reverse reactions are decreasing.
    (e) The rates of the forward and reverse reactions are equal.

11. Which of the following is true before a reaction reaches chemical equilibrium?
    (a) $rate_f$ = increase [reactant] / unit time
    (b) $rate_f$ = increase [product] / unit time
    (c) $rate_f$ = decrease [product] / unit time
    (d) all of the above
    (e) none of the above

Section 16.4 *General Equilibrium Constant, K_{eq}*

12. What is the equilibrium constant expression for: $3\,A + 2\,B \leftrightarrow 2\,C + D$?
    (a) $K_{eq} = [A]^3 [B]^2 / [C]^2 [D]^0$
    (b) $K_{eq} = [A]^3 [B]^2 / [C]^2 [D]^1$
    (c) $K_{eq} = [C]^2 [D]^0 / [A]^3 [B]^2$
    (d) $K_{eq} = [C]^2 [D]^1 / [A]^3 [B]^2$
    (e) none of the above

13. What is the equilibrium constant expression for:

    $$4\,HCl(g) + O_2(g) \leftrightarrow 2\,Cl_2(g) + 2\,H_2O(g)$$

    (a) $K_{eq} = [HCl] [O_2] / [Cl_2] [H_2O]$
    (b) $K_{eq} = [HCl]^4 [O_2] / [Cl_2]^2 [H_2O]^2$
    (c) $K_{eq} = [Cl_2] [H_2O] / [HCl] [O_2]$
    (d) $K_{eq} = [Cl_2]^2 [H_2O]^2 / [HCl]^4 [O_2]$
    (e) $K_{eq} = [H_2O]^2 / [HCl]^4$

14. What is the equilibrium constant expression for:

    $$2\,S(s) + 3\,O_2(g) \leftrightarrow 2\,SO_3(g)$$

    (a) $K_{eq} = [S] [O_2]^3 / [SO_3]^2$
    (b) $K_{eq} = [S] [O_2]^2 / [SO_3]^3$
    (c) $K_{eq} = [SO_3]^2 / [S] [O_2]^3$
    (d) $K_{eq} = [SO_3]^3 / [S] [O_2]^2$
    (e) $K_{eq} = [SO_3]^2 / [O_2]^3$

15. Oxygen and ozone gases are in chemical equilibrium in the upper atmosphere. Calculate K_{eq} for the reaction given the equilibrium concentrations of each gas at 20°C: $[O_2] = 9.38 \times 10^{-3}$ and $[O_3] = 3.40 \times 10^{-15}$.

$$2\,O_3(g) \quad \leftrightarrow \quad 3\,O_2(g)$$

(a) $K_{eq} = 4.11 \times 10^{-28}$
(b) $K_{eq} = 1.40 \times 10^{-23}$
(c) $K_{eq} = 3.62 \times 10^{-13}$
(d) $K_{eq} = 2.43 \times 10^{8}$
(e) $K_{eq} = 7.14 \times 10^{22}$

Section 16.5 *Gaseous State Equilibria Shifts*

16. Which of the changes listed below will shift the equilibrium to the right for the following reversible reaction?

$$4\,HCl(g) \;+\; O_2(g) \;+\; heat \quad \leftrightarrow \quad 2\,Cl_2(g) \;+\; 2\,H_2O(g)$$

(a) increase $[Cl_2]$
(b) decrease $[HCl]$
(c) decrease temperature
(d) increase pressure
(e) add a catalyst

17. Which of the changes listed below has no effect on the equilibrium for the following reversible reaction?

$$C(s) \;+\; O_2(g) \quad \leftrightarrow \quad CO_2(g) \;+\; heat$$

(a) increase $[CO_2]$
(b) increase $[O_2]$
(c) decrease $[O_2]$
(d) increase temperature
(e) increase pressure

Section 16.6 *Ionization Equilibrium Constant, K_i*

18. What is the equilibrium constant expression, K_i, for the weak acid H_2A?

$$H_2A(aq) \leftrightarrow H^+(aq) + HA^-(aq)$$

(a) $K_i = [H^+][HA^-]/[H_2A]$
(b) $K_i = [H^+]^2[HA^-]/[H_2A]$
(c) $K_i = [H^+]^2[A^-]/[H_2A]$
(d) $K_i = [H_2A]/[H^+][HA^-]$
(e) $K_i = [H_2A]/[H^+]^2[A^-]$

19. If the hydrogen ion concentration of 0.100 M chloroacetic acid, $HC_2H_2O_2Cl$, is 0.012 M, what is the ionization constant for the acid?

(a) $K_i = 1.2 \times 10^{-1}$
(b) $K_i = 1.2 \times 10^{-3}$
(c) $K_i = 1.4 \times 10^{-3}$
(d) $K_i = 1.4 \times 10^{-4}$
(e) $K_i = 1.4 \times 10^{-5}$

Section 16.7 *Weak Acid–Base Equilibria Shifts*

20. Which of the changes listed below will shift the equilibrium to the right for the following reversible reaction in aqueous solution?

$$HC_2H_3O_2(aq) \leftrightarrow H^+(aq) + C_2H_3O_2^-(aq)$$

(a) decrease pH
(b) decrease $[HC_2H_3O_2]$
(c) add solid $NaC_2H_3O_2$
(d) add solid NaOH
(e) none of the above

21. Which of the changes listed below will have no effect on the equilibrium for the following reversible reaction in aqueous solution?

$$NH_4OH(aq) \leftrightarrow NH_4^+(aq) + OH^-(aq)$$

(a) add solid NH_4Cl
(b) add solid KCl
(c) add gaseous HCl
(d) add gaseous NH_3
(e) all of the above

Section 16.8 *Solubility Product Equilibrium Constant,* K_{sp}

22. What is the equilibrium constant expression, K_{sp}, for slightly soluble silver phosphate in aqueous solution?

$$Ag_3PO_4(s) \leftrightarrow 3\,Ag^+(aq) + PO_4^{3-}(aq)$$

(a) $K_{sp} = [\,Ag^+\,]\,[\,PO_4^{3-}\,]$
(b) $K_{sp} = [\,Ag^+\,]\,[\,PO_4^{3-}\,]^3$
(c) $K_{sp} = [\,Ag^+\,]^3\,[\,PO_4^{3-}\,]$
(d) $K_{sp} = [\,Ag^+\,]^3\,[\,PO_4^{3-}\,] / [\,Ag_3PO_4\,]$
(e) $K_{sp} = [\,Ag^+\,]\,[\,PO_4^{3-}\,]^3 / [\,Ag_3PO_4\,]$

23. What is the K_{sp} for lead(II) fluoride, PbF_2, if the lead ion concentration in a saturated solution is 0.0019 M?

(a) $K_{sp} = 1.4 \times 10^{-5}$
(b) $K_{sp} = 7.2 \times 10^{-6}$
(c) $K_{sp} = 1.4 \times 10^{-8}$
(d) $K_{sp} = 2.7 \times 10^{-8}$
(e) $K_{sp} = 6.9 \times 10^{-9}$

Section 16.9 *Solubility Equilibria Shifts*

24. Which of the changes listed below will shift the equilibrium to the right for the following reversible reaction?

$$CaCO_3(s) \leftrightarrow Ca^{2+}(aq) + CO_3^{2-}(aq)$$

(a) increase $[\,Ca^{2+}\,]$
(b) increase $[\,CO_3^{2-}\,]$
(c) add solid $Ca(NO_3)_2$
(d) add solid $NaNO_3$
(e) decrease pH

25. Which of the changes listed below has no effect on the equilibrium for the following reversible reaction?

$$Hg_2I_2(s) \leftrightarrow Hg_2^{2+}(aq) + 2\,I^-(aq)$$

(a) increase $[\,Hg_2^{2+}\,]$
(b) increase $[\,I^-\,]$
(c) add solid Hg_2I_2
(d) add solid $Hg_2(NO_3)_2$
(e) add solid KI

Chapter 16 Key Terms

Across

1. an equilibrium with species in the same state
3. stated the effect of stress on a chemical equilibrium
4. energy of reactants – energy of products (3 words)
8. a chemical reaction that consumes heat energy
12. a general principle
14. a constant relating molarity of ions in solution
15. the speed that concentrations change (3 words)
16. a particle containing two or more nonmetal atoms
17. the state of highest energy on the reaction profile

Down

1. an equilibrium with species in different states
2. a chemical reaction that liberates heat energy
5. the energy required to reach the transition state
6. a reaction that proceeds toward reactants or products
7. a graph of the energy for a reaction
9. a substance that lowers the energy of activation
10. a dynamic state for a reversible reaction
11. a theory to explain the rate of a chemical reaction
13. an equilibrium constant for all reversible reactions

Chapter 16 Answers to Self–Test

1. **d** 2. **e** 3. **a** 4. **c** 5. **e** 6. **e** 7. **d** 8. **c** 9. **e** 10. **b** 11. **b** 12. **d** 13. **d** 14. **e** 15. **e**
16. **d** 17. **e** 18. **a** 19. **c** 20. **d** 21. **b** 22. **c** 23. **d** 24. **e** 25. **c**

Chapter 16 Answers to Crossword Puzzle

Across
1. homogeneous
3. LeChatelier
4. heat of reaction
8. endothermic
12. law
14. solubility
15. rate of reaction
16. molecule
17. transition

Down
1. heterogeneous
2. exothermic
5. activation
6. reversible
7. profile
9. catalyst
10. equilibrium
11. collision
13. general

Oxidation and Reduction

Section 17.1 *Oxidation Numbers*

1. What is the oxidation number of metallic iron wire?
   (a) 0
   (b) +2
   (c) +3
   (d) all of the above
   (e) none of the above

2. What is the oxidation number of greenish-yellow chlorine gas in the free state?
   (a) 0   (b) −1   (c) +1   (d) +3   (e) +5

3. What is the oxidation number of mercury in the mercurous ion, Hg_2^{2+} ?
   (a) 0
   (b) +1
   (c) +2
   (d) +4
   (e) none of the above

4. What is the oxidation number of chromium in the dichromate ion, $Cr_2O_7^{2-}$?
   (a) −2
   (b) +2
   (c) +6
   (d) +7
   (e) none of the above

5. What is the oxidation number of carbon in $Pb(HCO_3)_4$?
   (a) −4
   (b) +2
   (c) +3
   (d) +4
   (e) none of the above

6. What is the oxidation number of chlorine in aqueous $HClO_3$?
   (a) −1
   (b) +1
   (c) +3
   (d) +5
   (e) none of the above

Section 17.2 *Oxidation–Reduction Reactions*

7. What substance is oxidized in the following redox reaction?

$$Cl_2(g) \quad + \quad 2\,Br^-(aq) \quad \rightarrow \quad 2\,Cl^-(aq) \quad + \quad Br_2(l)$$

   (a) Cl_2
   (b) Br^-
   (c) Cl^-
   (d) Br_2
   (e) H_2O

8. What substance is the oxidizing agent in the preceding redox reaction?
   (a) Cl_2
   (b) Br^-
   (c) Cl^-
   (d) Br_2
   (e) H_2O

Section 17.3 *Balancing Redox Equations: Oxidation Number Method*

9. After balancing the following redox reaction, what is the coefficient of Mn?

$$Fe_2O_3(l) \quad + \quad Mn(l) \quad \rightarrow \quad MnO_2(l) \quad + \quad Fe(l)$$

   (a) 1
   (b) 2
   (c) 3
   (d) 4
   (e) none of the above

10. After balancing the following redox reaction, what is the coefficient of H^+?

$$MnO_4^-(aq) + VO_2^-(aq) + H^+(aq) \quad \rightarrow \quad Mn^{2+}(aq) + VO_3^-(aq) + H_2O(l)$$

   (a) 2
   (b) 3
   (c) 5
   (d) 6
   (e) none of the above

11. Which of the following is *not* a guideline for balancing redox equations by the oxidation number method?
    (a) Diagram the electrons lost by the substance oxidized and gained by the substance reduced.
    (b) Write a half–reaction for the substance oxidized and the substance reduced.
    (c) Place a coefficient in front of the substance oxidized that corresponds to the number of electrons gained by the substance reduced.
    (d) Place a coefficient in front of the substance reduced that corresponds to the number of electrons lost by the substance oxidized.
    (e) Verify the total number of atoms and total ionic charge is equal for reactants and products.

Section 17.4 *Balancing Redox Equations: Half–Reaction Method*

12. Which of the following is *not* a guideline for balancing redox equations in acid by the half–reaction method?
    (a) Write a half–reaction for the substance oxidized and the substance reduced.
    (b) Balance the atoms in each half–reaction; balance oxygen with water and hydrogen with H^+.
    (c) Multiply each half–reaction by a whole number so that the number of electrons lost by the substance oxidized is equal to the electrons gained by the substance reduced.
    (d) Add the two half–reactions together and cancel identical species on each side of the equation.
    (e) Verify that the total number of atoms is equal to the total ionic charge.

13. Which of the following is *not* a guideline for balancing redox equations in base by the half–reaction method?
    (a) Write a half–reaction for the substance oxidized and the substance reduced.
    (b) Balance the atoms in each half–reaction; balance oxygen with water and hydrogen with OH^-.
    (c) Multiply each half–reaction by a whole number so that the number of electrons lost by the substance oxidized is equal to the electrons gained by the substance reduced.
    (d) Add the two half–reactions together and cancel identical species on each side of the equation.
    (e) Verify that the total ionic charge is equal for reactants and products.

14. After balancing the following redox reaction in acidic solution, what is the coefficient of water?

$$MnO_4^-(aq) \ + \ Co^{2+}(aq) \quad \rightarrow \quad Mn^{2+}(aq) \ + \ Co^{3+}(aq)$$

(a) 2
(b) 3
(c) 4
(d) 5
(e) none of the above

15. After balancing the following redox reaction in basic solution, what is the coefficient of water?

$$MnO_4^-(aq) \ + \ Co^{2+}(aq) \quad \rightarrow \quad MnO_2(s) \ + \ Co^{3+}(aq)$$

(a) 2
(b) 3
(c) 5
(d) 6
(e) none of the above

Section 17.5 *Predicting Spontaneous Redox Reactions*

16. Which of the substances listed below is the *strongest* oxidizing agent given the following *spontaneous* redox reaction?

$$FeCl_3(aq) \ + \ NaI(aq) \quad \rightarrow \quad I_2(s) \ + \ FeCl_2(aq) \ + \ NaCl(aq)$$

(a) FeCl$_3$
(b) NaI
(c) I$_2$
(d) FeCl$_2$
(e) NaCl

17. Which of the substances listed below is the *strongest* reducing agent given the following *nonspontaneous* redox reaction?

$$FeCl_3(aq) \ + \ NaBr(aq) \quad \rightarrow \quad Br_2(l) \ + \ FeCl_2(aq) \ + \ NaCl(aq)$$

(a) FeCl$_3$
(b) NaBr
(c) Br$_2$
(d) FeCl$_2$
(e) NaCl

18. Given that the following redox reactions go essentially to completion, which of the metals listed below has the *greatest* tendency to undergo oxidation?

$$Co(s) + Sn^{2+}(aq) \rightarrow Sn(s) + Co^{2+}(aq)$$
$$Al(s) + Zn^{2+}(aq) \rightarrow Zn(s) + Al^{3+}(aq)$$
$$Sn(s) + Ag^{+}(aq) \rightarrow Ag(s) + Sn^{2+}(aq)$$
$$Zn(s) + Co^{2+}(aq) \rightarrow Co(s) + Zn^{2+}(aq)$$

(a) Ag
(b) Co
(c) Sn
(d) Al
(e) Zn

19. Given that the above redox reactions go essentially to completion, which of the metals listed below is the *strongest* oxidizing agent?

(a) Ag
(b) Co
(c) Sn
(d) Al
(e) Zn

Section 17.6 *Voltaic Cells*

20. Which of the statements listed below is true regarding the following redox reaction occurring in a *spontaneous* electrochemical cell?

$$Cr(s) + Ni^{2+}(aq) \rightarrow Ni(s) + Cr^{3+}(aq)$$

(a) Cr is oxidized at the anode
(b) Ni^{2+} is reduced at the anode
(c) electrons flow from the Ni electrode to the Cr electrode
(d) anions in the salt bridge flow from the Cr half-cell to the Ni half-cell
(e) all of the above

21. Which of the statements listed below is true regarding the following redox reaction occurring in a *spontaneous* electrochemical cell?

$$F_2(g) + 2\,Br^-(aq) \rightarrow Br_2(l) + 2\,F^-(aq)$$

(a) Br^- is oxidized at the anode
(b) F_2 is reduced at the cathode
(c) electrons flow from the anode to the cathode
(d) anions in the salt bridge flow from the F_2 half-cell to the Br_2 half-cell
(e) all of the above

22. Nickel-cadmium batteries are used in rechargeable electronic calculators. Given the *spontaneous* reaction for a discharging nicad battery, what substance is being *reduced*?

$$Cd(s) + NiO_2(s) + 2 H_2O(l) \rightarrow Cd(OH)_2(s) + Ni(OH)_2(s)$$

(a) Cd
(b) NiO_2
(c) H_2O
(d) $Cd(OH)_2$
(e) $Ni(OH)_2$

23. Which of the following is an example of a spontaneous electrochemical cell that has a commercial application as a common voltaic cell?
(a) alkaline dry cell
(b) mercury dry cell
(c) lead storage battery
(d) nicad battery
(e) all of the above

Section 17.7 *Electrolytic Cells*

24. Which of the statements listed below is true regarding the following redox reaction occurring in a *nonspontaneous* electrochemical cell?

$$MgCl_2(l) \xrightarrow{\text{electricity}} Mg(l) + Cl_2(g)$$

(a) magnesium metal is produced at the anode
(b) chlorine gas is produced at the anode
(c) oxidation half-reaction: $2 Cl^- \rightarrow Cl_2 + 2 e^-$
(d) reduction half-reaction: $Mg^{2+} + 2 e^- \rightarrow Mg$
(e) none of the above

25. Which of the statements listed below is true regarding the following redox reaction in a *nonspontaneous* electrochemical cell?

$$Br_2(l) + 2 NaCl(aq) \xrightarrow{\text{electricity}} Cl_2(g) + 2 NaBr(aq)$$

(a) liquid bromine is oxidized at the anode
(b) gaseous chlorine is reduced at the cathode
(c) electrons flow from the anode to the cathode
(d) bromine is a stronger oxidizing agent than chlorine
(e) none of the above

Chapter 17 Key Terms

Across

1. the agent undergoing oxidation
4. a cell having two electrodes connected by a wire
7. the agent undergoing reduction
8. a nonspontaneous electrochemical cell
10. the conversion of chemical and electrical energy
12. indicates an electron rich or poor atom (2 words)
15. the electrode at which oxidation occurs
16. an electrochemical cell without aqueous solution
18. the ability of a substance to undergo reduction

Down

2. the electrode at which reduction occurs
3. a cell having a single electrode
5. a process in which a substance gains electrons
6. a spontaneous electrochemical cell
9. allows ions to move between half-cells (2 words)
11. a process in which a substance loses electrons
13. a term for any device that produces electricity
14. a reaction in which electron transfer occurs
17. a reaction involving either oxidation or reduction

Chapter 17 Answers to Self–Test

1. a 2. a 3. b 4. c 5. d 6. d 7. b 8. a 9. c 10. d 11. b 12. e 13. b 14. c 15. a
16. a 17. d 18. d 19. a 20. a 21. e 22. b 23. e 24. b 25. c

Chapter 17 Answers to Crossword Puzzle

Across
1. reducing
4. electrochemical
7. oxidizing
8. electrolytic
10. electrochemistry
12. oxidation number
15. anode
16. dry
18. potential

Down
2. cathode
3. half-cell
5. reduction
6. voltaic
9. salt bridge
11. oxidation
13. battery
14. redox
17. half

Nuclear Chemistry

CHAPTER

18

Section 18.1 *Natural Radioactivity*

1. What are the three main types of radioactive emission from an atomic nucleus?
   - (a) proton, neutron, and electron
   - (b) proton, neutron, and positron
   - (c) positron, neutron, and electron
   - (d) positron, neutron, and electron
   - (e) alpha, beta, and gamma

2. Which type of nuclear radiation is identical to a helium nucleus and is deflected toward the negative electrode as it passes between electrically charged plates?
   - (a) alpha
   - (b) beta
   - (c) gamma
   - (d) all of the above
   - (e) none of the above

3. Which type of natural radiation can pass through the human body and requires thick lead protective shielding?
   - (a) alpha
   - (b) beta
   - (c) gamma
   - (d) all of the above
   - (e) none of the above

Section 18.2 *Nuclear Equations*

4. What is the name of the particle having the following atomic notation: $^{4}_{2}\text{He}$?
   - (a) alpha
   - (b) beta
   - (c) gamma
   - (d) neutron
   - (e) positron

5. What is the name of the radiation having the following atomic notation: $^0_0\gamma$?
   - (a) alpha
   - (b) beta
   - (c) gamma
   - (d) neutron
   - (e) positron

6. What is the name of the particle having the following atomic notation: $^1_0 n$?
   - (a) alpha
   - (b) electron
   - (c) neutron
   - (d) positron
   - (e) proton

7. What is the approximate mass and relative charge of a beta particle?
   - (a) 4 amu and 2+
   - (b) 1 amu and 1+
   - (c) 1 amu and 0
   - (d) 0 amu and 1–
   - (e) 0 amu and 1+

8. What is the approximate mass and relative charge of a positron?
   - (a) 4 amu and 2+
   - (b) 1 amu and 1+
   - (c) 1 amu and 0
   - (d) 0 amu and 1–
   - (e) 0 amu and 1+

9. What is the approximate mass and relative charge of a proton?
   - (a) 4 amu and 2+
   - (b) 1 amu and 1+
   - (c) 1 amu and 0
   - (d) 0 amu and 1–
   - (e) 0 amu and 1+

10. What nuclide is produced when a K–43 nucleus decays by beta emission?

    (a) $^{43}_{18}Ar$ (b) $^{42}_{19}K$ (c) $^{42}_{20}Ca$ (d) $^{43}_{20}Ca$ (e) $^{44}_{20}Ca$

11. What nuclide is produced when an O–15 nucleus decays by positron emission?

    (a) $^{14}_{6}C$ (b) $^{14}_{7}N$ (c) $^{15}_{7}N$ (d) $^{14}_{8}O$ (e) $^{15}_{9}F$

12. What nuclide is produced when a Ba–133 nucleus decays by electron capture?

(a) $^{132}_{55}Cs$ (b) $^{133}_{55}Cs$ (c) $^{132}_{56}Ba$ (d) $^{132}_{57}La$ (e) $^{133}_{57}La$

Section 18.3 *Radioactive Decay Series*

13. The uranium–238 decay series begins with the emission of an alpha particle. If the daughter decays by beta emission, what is the resulting particle?

(a) $^{234}_{90}Th$
(b) $^{233}_{91}Pa$
(c) $^{234}_{91}Pa$
(d) $^{234}_{93}Np$
(e) none of the above

14. In the final step of the uranium–238 disintegration series, the parent nuclide decays into lead–206 and an alpha particle. What is the parent nuclide?

(a) $^{202}_{80}Hg$
(b) $^{210}_{83}Bi$
(c) $^{206}_{84}Po$
(d) $^{210}_{84}Po$
(e) none of the above

Section 18.4 *Radioactive Half–Life*

15. Each "click" registered by a Geiger counter indicates which of the following?
(a) one half-life
(b) one scintillation
(c) one minute elapsing
(d) one second elapsing
(e) one nucleus decaying

16. The initial activity of a radionuclide was 1000 dpm and dropped to 125 dpm after 72 days. What is the half-life of the radionuclide?
(a) $t_{1/2}$ = 18 days
(b) $t_{1/2}$ = 24 days
(c) $t_{1/2}$ = 36 days
(d) $t_{1/2}$ = 144 days
(e) $t_{1/2}$ = 216 days

17. If 100 mg of a radionuclide is used for medical diagnosis, how much of the radionuclide is still radioactive after 30 hours? $(t_{1/2} = 15\ hours)$
    (a) 25 mg
    (b) 33 mg
    (c) 50 mg
    (d) 100 mg
    (e) 300 mg

18. If the initial activity of a radionuclide is 160 dpm, how much time elapses before the activity drops to 10 dpm? $(t_{1/2} = 48\ minutes)$
    (a) 12 min
    (b) 96 min
    (c) 144 min
    (d) 192 min
    (e) 768 min

Section 18.5 *Radionuclide Applications*

19. An archaeologist discovers a parchment with a carbon–14 activity of 7 dpm per gram of carbon. If carbon–14 has an activity of 14 dpm, what is the estimated age of the parchment? $(t_{1/2} = 5730\ years)$
    (a) 1400 years
    (b) 2900 years
    (c) 5700 years
    (d) 11,000 years
    (e) 17,000 years

20. What radionuclide can be used to irradiate food and destroy microorganisms?
    (a) cobalt–60
    (b) iodine–131
    (c) iridium–192
    (d) sodium–24
    (e) xenon–133

Section 18.6 *Artificial Radioactivity*

21. Bombarding sodium–23 with a proton produces particle X and a neutron. What is particle X ?
    (a) neon–23
    (b) sodium–24
    (c) magnesium–23
    (d) magnesium–24
    (e) none of the above

22. Firing an accelerated particle at a boron–10 target nucleus produces nitrogen–14 and a gamma ray. What is the projectile particle?
    (a) alpha
    (b) beta
    (c) neutron
    (d) proton
    (e) positron

Section 18.7 *Nuclear Fission*

23. A single neutron causes uranium–235 to fission and release three neutrons. Assuming each of the neutrons causes a second fission that releases two neutrons, how many neutrons are released in the second step?
    (a) 3
    (b) 6
    (c) 8
    (d) 9
    (e) none of the above

24. How many neutrons are produced from the following fission reaction?

$$_{92}^{235}U + \,_{0}^{1}n \rightarrow \,_{52}^{137}Te + \,_{40}^{96}Zr + \,_{0}^{1}n$$

    (a) 1
    (b) 2
    (c) 3
    (d) 4
    (e) none of the above

Section 18.8 *Mass Defect and Binding Energy*

25. If a helium–3 nucleus has a mass of 3.0149 amu, what is the mass defect? (The mass of a proton is 1.0073 amu and a neutron is 1.0087.)
    (a) 0.0070 amu
    (b) 0.0084 amu
    (c) 0.0098 amu
    (d) 0.0112 amu
    (e) 0.9989 amu

26. If a beryllium–8 nucleus has a mass defect of 0.0587 g/mol, what is the binding energy in J/mol?
    (a) 1.43×10^4 J/mol
    (b) 1.43×10^7 J/mol
    (c) 3.91×10^9 J/mol
    (d) 5.28×10^{12} J/mol
    (e) 5.28×10^{15} J/mol

Section 18.9 *Nuclear Fusion*

27. Which of the following best describes the process of nuclear fusion?
    (a) Combining two small nuclei into a larger nucleus at 25°C.
    (b) Combining two small nuclei into a larger nucleus at ~10,000,000°C.
    (c) Combining two small nuclei into a larger nucleus at ~1000 K.
    (d) Splitting a large nucleus into two smaller nuclei at 25°C.
    (e) Splitting a large nucleus into two smaller nuclei at ~10,000,000°C.

28. The nuclear fusion of deuterium and tritium gives a neutron and particle–X. Identify particle–X.
    (a) hydrogen–3
    (b) helium–3
    (c) helium–4
    (d) lithium–5
    (e) none of the above

Nuclear Equations for Transmutation Reactions

29. Phosphorus–30 was the first synthetic radionuclide and was produced by bombarding aluminum–27 with an alpha particle. Write a balanced nuclear equation for the reaction.

30. Element 106 was first synthesized in Germany by firing an oxygen–18 projectile at a californium–249 target. Write a balanced nuclear equation for the reaction if the nuclide produced had a mass number of 263.

Chapter 18 Key Terms

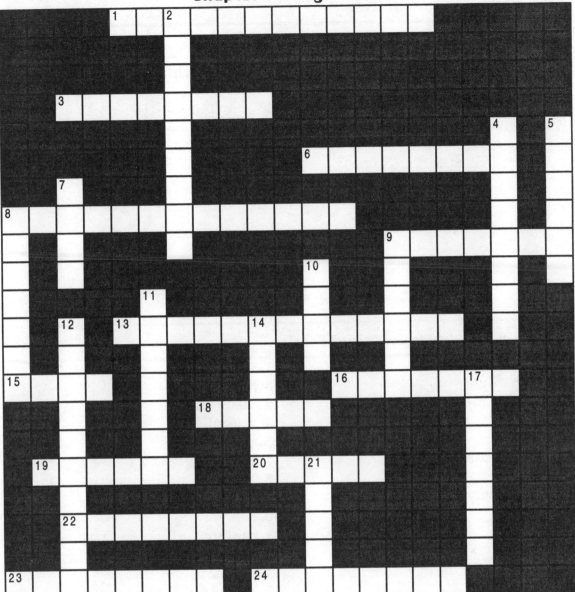

Across
1. a nucleus of a specific radioactive isotope
3. a positive electron
6. the nuclide resulting from a decaying nucleus
8. the conversion of one element into another
9. a reaction in which a nucleus splits
13. the emission of particles from an unstable nucleus
15. the A value refers to the _____ number
16. the energy holding a nucleus together
18. a nuclear radiation identical to high-energy light
19. the stepwise disintegration of a radionuclide
20. a fission reaction that initiates additional fissions
22. accounts for protons and neutrons in a reaction
23. the rate of decay of radionuclides
24. a description of a change involving a nucleus

Down
2. the nuclide of hydrogen having one neutron
4. the minimum mass that sustains a chain reaction
5. a nuclide that decays to give a daughter product
7. the life of 50% of the radionuclides in a sample
8. the nuclide of hydrogen having two neutrons
9. a reaction in which two small nuclei combine
10. a nuclear radiation identical to an electron
11. a heavy nuclide draws an electron into its nucleus
12. the mass equivalent of binding energy (2 words)
14. the Z value refers to the _____ number
17. a nucleus of a specific isotope
21. a nuclear radiation identical to a helium-4 nucleus

Chapter 18 Answers to Self–Test

1. **e** 2. **a** 3. **c** 4. **a** 5. **c** 6. **c** 7. **d** 8. **e** 9. **b** 10. **d** 11. **c** 12. **b** 13. **c** 14. **d** 15. **e**
16. **b** 17. **a** 18. **d** 19. **c** 20. **a** 21. **c** 22. **a** 23. **b** 24. **c** 25. **b** 26. **d** 27. **b** 28. **c**

Nuclear Equations for Transmutation Reactions

29.
$$^{27}_{13}\text{Al} \quad + \quad ^{4}_{2}\text{He} \quad \rightarrow \quad ^{30}_{15}\text{P} \quad + \quad ^{1}_{0}\text{n}$$

30.
$$^{249}_{98}\text{Cf} \quad + \quad ^{18}_{8}\text{O} \quad \rightarrow \quad ^{263}_{106}\text{Sg} \quad + \quad 4\,^{1}_{0}\text{n}$$

Chapter 18 Answers to Crossword Puzzle

Across
1. radionuclide
3. positron
6. daughter
8. transmutation
9. fission
13. radioactivity
15. mass
16. binding
18. gamma
19. series
20. chain
22. equation
23. activity
24. reaction

Down
2. deuterium
4. critical
5. parent
7. half
8. tritium
9. fusion
10. beta
11. capture
12. mass defect
14. atomic
17. nuclide
21. alpha

Organic Chemistry

Section 19.1 *Hydrocarbons*

1. Which of the following is a class of saturated hydrocarbons?
   (a) alkanes
   (b) alkenes
   (c) alkynes
   (d) aromatic
   (e) none of the above

Section 19.2 *Alkanes*

2. What is the IUPAC name of the following alkane:

   $$CH_3-CH_2-CH_2-CH_2-CH_2-CH_2-CH_2-CH_2-CH_3 ?$$

   (a) decane
   (b) heptane
   (c) hexane
   (d) nonane
   (e) none of the above

3. How many isomers share the molecular formula C_6H_{14}?

   (a) 2   (b) 3   (c) 4   (d) 5   (e) 6

4. What is the name of the following alkyl group: $(CH_3)_2CH-$?
   (a) butyl
   (b) ethyl
   (c) isopropyl
   (d) methyl
   (e) propyl

5. Which of the following is the condensed structural formula for 2–methylhexane?
   (a) $CH_3–CH(CH_3)–CH_2–CH_2–CH_3$
   (b) $CH_3–CH(CH_3)–CH_2–CH_2–CH_2–CH_3$
   (c) $CH_3–CH_2–CH(CH_3)–CH_2–CH_2–CH_3$
   (d) $CH_3–CH(CH_3)–CH_2–CH_2–CH_2–CH_2–CH_3$
   (e) $CH_3–CH_2–CH(CH_3)–CH_2–CH_2–CH_2–CH_3$

6. What are the products from the complete combustion of propane gas?
   (a) CH_4 and H_2O
   (b) CO and H_2O
   (c) CO and H_2
   (d) CO_2 and H_2O
   (e) CO_2 and H_2

Section 19.3 *Alkenes and Alkynes*

7. Which of the following is the condensed structural formula for 3–octene?
   (a) $CH_3–CH_2–CH=CH–CH_2–CH_2–CH_3$
   (b) $CH_3–CH_2–CH_2–CH=CH–CH_2–CH_3$
   (c) $CH_3–CH_2–CH_2–CH_2–CH=CH–CH_2–CH_3$
   (d) $CH_3–CH_2–CH_2–CH=CH–CH_2–CH_2–CH_3$
   (e) none of the above

8. What is the IUPAC name of the following: $CH_3–CH_2–C{\equiv}CH$?
   (a) butyne
   (b) 1–butyne
   (c) 2–butyne
   (d) 3–butyne
   (e) none of the above

9. What are the products from the complete combustion of acetylene?
   (a) C_2H_2 and H_2O
   (b) CO and H_2O
   (c) CO and H_2
   (d) CO_2 and H_2O
   (e) CO_2 and H_2

10. Which of the following illustrates the hydrogenation of propylene gas with nickel catalyst at 25°C and 1 atm?

(a) $CH_2=CH-CH_3 + H_2 \xrightarrow{Ni} C_3H_4$

(b) $CH_2=CH-CH_3 + H_2 \xrightarrow{Ni} C_3H_6$

(c) $CH_2=CH-CH_3 + H_2 \xrightarrow{Ni} C_3H_8$

(d) $CH_2=CH-CH_3 + H_2 \xrightarrow{Ni} C_6H_6$

(e) $CH_2=CH-CH_3 + H_2 \xrightarrow{Ni}$ no reaction

Section 19.4 *Aromatic Hydrocarbons*

11. How many representations of benzene are there according to the delocalized electron model?
    (a) 1
    (b) 2
    (c) 3
    (d) 4
    (e) none of the above

12. How many different isomers are there for dimethylbenzene, $C_6H_4(CH_3)_2$?
    (a) 0
    (b) 2
    (c) 3
    (d) 4
    (e) none of the above

Section 19.5 *Hydrocarbon Derivatives*

13. Which of the following classes of compounds does not have a carbonyl group?
    (a) aldehyde
    (b) amide
    (c) ester
    (d) ketone
    (e) phenol

14. What is the general formula for an organic halide?
    (a) $R-X$
    (b) $R-OH$
    (c) $Ar-OH$
    (d) $R-O-R'$
    (e) $R-NH_2$

15. What is the general formula for an alcohol?
    (a) R—X
    (b) R—OH
    (c) Ar—OH
    (d) R—O—R'
    (e) R—NH$_2$

16. What is the general formula for an ether?
    (a) R—X
    (b) R—OH
    (c) Ar—OH
    (d) R—O—R'
    (e) R—NH$_2$

17. What class of compound has the following general formula: $\overset{\overset{\displaystyle O}{\|}}{R—C—H}$?
    (a) aldehyde
    (b) ketone
    (c) carboxylic acid
    (d) ester
    (e) amide

18. What class of compound has the following general formula: $\overset{\overset{\displaystyle O}{\|}}{R—C—OH}$?
    (a) aldehyde
    (b) ketone
    (c) carboxylic acid
    (d) ester
    (e) amide

19. What class of compound is the following: C$_6$H$_5$–OH?
    (a) organic halide
    (b) alcohol
    (c) phenol
    (d) ether
    (e) amine

20. What class of compound is the following: CH$_3$CH$_2$–NH$_2$?
    (a) organic halide
    (b) alcohol
    (c) phenol
    (d) ether
    (e) amine

Section 19.6 *Organic Halides*

21. What is the condensed structural formula for ethyl bromide?
    (a) CH_3-Br
    (b) CH_3-CH_2-Br
    (c) $CH_3-CH_2-CH_2-Br$
    (d) $CH_3-CHBr-CH_3$
    (e) none of the above

Section 19.7 *Alcohols, Phenols, and Ethers*

22. What is the condensed structural formula for isopropyl alcohol?
    (a) CH_3-CH_2-OH
    (b) $CH_3-CH_2-CH_2-OH$
    (c) $CH_3-CH(OH)-CH_3$
    (d) $CH_3-CH(OH)-CH_2-CH_3$
    (e) none of the above

23. What is the common name for the following: $CH_3CH_2-O-CH_3$?
    (a) diethyl ether
    (b) dimethyl ether
    (c) ethyl ether
    (d) methyl ether
    (e) ethyl methyl ether

Section 19.8 *Amines*

24. What is the condensed structural formula for propyl amine?
    (a) $CH_3-CH_2-NH_2$
    (b) $CH_3-CH_2-CH_2-NH_2$
    (c) $CH_3-CH(NH_2)-CH_3$
    (d) $CH_3-CH(NH_2)-CH_2-CH_3$
    (e) none of the above

Section 19.9 *Aldehydes and Ketones*

25. Which of the following organic compounds is most likely an aldehyde based
    upon the nomenclature suffix?
    (a) chlordane (insecticide)
    (b) cholesterol (steroid)
    (c) citral (lemon fragrance)
    (d) codeine (narcotic)
    (e) cresol (antiseptic)

26. What is the IUPAC systematic name for the following: $CH_3-\overset{\overset{\displaystyle O}{\|}}{C}-CH_3$?
   (a) acetone
   (b) dimethyl ketone
   (c) propanaldehyde
   (d) propanal
   (e) propanone

Section 19.10 *Carboxylic Acids, Esters, and Amides*

27. What is the IUPAC systematic name for the following: $CH_3-\overset{\overset{\displaystyle O}{\|}}{C}-OH$?
   (a) acetic acid
   (b) ethanoic acid
   (c) formic acid
   (d) methanoic acid
   (e) propanoic acid

28. What is the IUPAC systematic name for the following: $H-\overset{\overset{\displaystyle O}{\|}}{C}-O-CH_2CH_3$?
   (a) ethyl formate
   (b) ethyl methanoate
   (c) ethyl propanoate
   (d) methyl acetate
   (e) methyl ethanoate

IUPAC Nomenclature of Organic Compounds

29. What is the IUPAC name for the following compound?

$$CH_3\text{--}CH_2\text{--}CH\text{--}CH_2\text{--}CH\text{--}CH_3$$
$$\underset{CH_3}{|} \qquad \underset{CH_2\text{--}CH_3}{|}$$

30. What is the IUPAC name for the following compound?

$$CH_2\text{=}CH\text{--}CH_2\text{--}CH\text{--}CH_2\text{--}CH_3$$
$$\underset{CH_2\text{--}CH_3}{|}$$

Chapter 19 Key Terms

Across

1. a family of saturated hydrocarbons
3. a compound containing C, H, and O
5. a hydrocarbon containing all single bonds
9. a carbon and oxygen double bond
10. the second member of the alkane family
11. a compound containing only hydrogen and carbon
16. the same molecular formula but different structures
18. a formula showing the molecular arrangement
19. an alkane fragment after removing a H atom
20. a family of compounds
21. the first member of the alkane family

Down

1. a reaction of a hydrocarbon with H_2 or Br_2
2. a group typical of a class of compounds
4. a family of compounds that are aromatic
6. an arene fragment after removing a H atom
7. the functional group in a carboxylic acid
8. a benzene fragment after removing a H atom
9. a reaction of a hydrocarbon with O_2
11. the functional group in an alcohol or phenol
12. the compounds containing carbon
13. a family of compounds with a double bond
14. a hydrocarbon containing double or triple bonds
15. a hydrocarbon containing a benzene ring
17. a family of compounds with a triple bond

Chapter 19 Answers to Self–Test

1. **a** 2. **d** 3. **d** 4. **c** 5. **b** 6. **d** 7. **c** 8. **b** 9. **d** 10. **c** 11. **a** 12. **c** 13. **e** 14. **a** 15. **b** 16. **d**
17. **a** 18. **c** 19. **c** 20. **e** 21. **b** 22. **c** 23. **e** 24. **b** 25. **c** 26. **e** 27. **b** 28. **b**

IUPAC Nomenclature of Organic Compounds

29. 3,5–dimethylheptane

30. 4–ethyl–1–hexene

Chapter 19 Answers to Crossword Puzzle

Across
1. alkanes
3. derivative
5. saturated
9. carbonyl
10. ethane
11. hydrocarbon
16. isomers
18. structural
19. alkyl
20. class
21. methane

Down
1. addition
2. functional
4. arenes
6. aryl
7. carboxyl
8. phenyl
9. combustion
11. hydroxyl
12. organic
13. alkenes
14. unsaturated
15. aromatic
17. alkynes

Biochemistry

Section 20.1 *Biological Compounds*

1. What type of biological compound is a polymer of amino acids?
   - (a) protein
   - (b) carbohydrate
   - (c) lipid
   - (d) nucleic acid
   - (e) none of the above

2. What type of biological compound is a polymer of simple sugar molecules?
   - (a) protein
   - (b) carbohydrate
   - (c) lipid
   - (d) nucleic acid
   - (e) none of the above

3. What type of biological compound is characterized by carboxylic acid, alcohol, and ester functional groups?
   - (a) protein
   - (b) carbohydrate
   - (c) lipid
   - (d) nucleic acid
   - (e) none of the above

4. What type of biological compound is characterized by aldehyde, alcohol, and amine functional groups?
   - (a) protein
   - (b) carbohydrate
   - (c) lipid
   - (d) nucleic acid
   - (e) none of the above

5. Which of the following types of bonds is found in a protein?
   (a) peptide linkage
   (b) glycoside linkage
   (c) ester linkage
   (d) phosphate linkage
   (e) none of the above

6. Which of the following types of bonds is found in a carbohydrate?
   (a) peptide linkage
   (b) glycoside linkage
   (c) ester linkage
   (d) phosphate linkage
   (e) none of the above

7. Which of the following types of bonds is found in a lipid?
   (a) peptide linkage
   (b) glycoside linkage
   (c) ester linkage
   (d) phosphate linkage
   (e) none of the above

8. Which of the following types of bonds is found in a nucleic acid?
   (a) peptide linkage
   (b) glycoside linkage
   (c) ester linkage
   (d) phosphate linkage
   (e) none of the above

Section 20.2 *Proteins*

9. What is the overall shape of a protein that serves a structural function?
   (a) α–helix
   (b) globular
   (c) long and extended
   (d) pleated sheet
   (e) none of the above

10. What is the overall shape of a protein that serves a metabolic function?
   (a) α–helix
   (b) globular
   (c) long and extended
   (d) pleated sheet
   (e) none of the above

11. What is the overall shape of a protein after it undergoes denaturation?
    (a) α–helix
    (b) globular
    (c) long and extended
    (d) pleated sheet
    (e) random shape

12. What type of bonds are responsible for the secondary structure of a protein?
    (a) amide bonds
    (b) ester bonds
    (c) hydrogen bonds
    (d) ionic bonds
    (e) none of the above

13. What type of protein structure corresponds to a pleated sheet of amino acids?
    (a) primary
    (b) secondary
    (c) tertiary
    (d) quaternary
    (e) none of the above

14. How many amino acids compose valylalanyltyrosine?
    (a) 1
    (b) 2
    (c) 3
    (d) 4
    (e) 1000s

15. How many amino acid sequences are possible for a tripeptide containing proline (pro), histidine (his), and leucine (leu)?
    (a) 1
    (b) 3
    (c) 6
    (d) 9
    (e) thousands

Section 20.3 *Enzymes*

16. An enzyme is an example of which type of biological compound?
    (a) protein
    (b) carbohydrate
    (c) lipid
    (d) nucleic acid
    (e) none of the above

17. Which of the following occurs in Step 1 of enzyme catalysis?
    (a) The substrate cleaves the enzyme molecule.
    (b) The enzyme cleaves the substrate molecule.
    (c) The substrate molecule binds to the active site.
    (d) The enzyme molecule binds to the active site.
    (e) none of the above

18. Which of the following occurs in Step 2 of enzyme catalysis?
    (a) The substrate cleaves the enzyme molecule.
    (b) The enzyme cleaves the substrate molecule.
    (c) The substrate molecule binds to the active site.
    (d) The enzyme molecule binds to the active site.
    (e) none of the above

19. What is the term for a molecule that blocks the active site on an enzyme?
    (a) catalyst
    (b) inhibitor
    (c) steroid
    (d) substrate
    (e) none of the above

Section 20.4 *Carbohydrates*

20. Which of the following functional groups are found in a ketopentose?
    (a) alcohol and aldehyde
    (b) alcohol and ketone
    (c) aldehyde and ketone
    (d) aldehyde and phenol
    (e) none of the above

21. How many carbon atoms are in a molecule of ketopentose sugar?
    (a) 1
    (b) 5
    (c) 6
    (d) 12
    (e) none of the above

22. Which monosaccharides are produced from the acid hydrolysis of sucrose?
    (a) fructose and glucose
    (b) galactose and glucose
    (c) gulose and glucose
    (d) ribose and glucose
    (e) none of the above

23. What is the repeating monosaccharide in cellulose?
    (a) fructose
    (b) galactose
    (c) glucose
    (d) ribose
    (e) none of the above

Section 20.5 *Lipids*

24. Which of the following are characteristic of a lipid fat?
    (a) liquid
    (b) plant source
    (c) unsaturated fatty acids
    (d) water–insoluble
    (e) all of the above

25. Which of the following is not found in a fat or oil?
    (a) ester linkages
    (b) glycerol
    (c) saturated fatty acids
    (d) steroid
    (e) unsaturated fatty acids

26. Which of the following is the structure of palmitic acid?
    (a) $CH_3-(CH_2)_{10}-COOH$
    (b) $CH_3-(CH_2)_{12}-COOH$
    (c) $CH_3-(CH_2)_{14}-COOH$
    (d) $CH_3-(CH_2)_{16}-COOH$
    (e) none of the above

27. What are the products from the saponification of trilinolein?
    (a) water and sodium linoleate
    (b) water and linoleic acid
    (c) glycerol and linoleic acid
    (d) glycerol and sodium linoleate
    (e) none of the above

28. What are the products from the saponification of a lipid wax?
    (a) long–chain alcohol and a fatty acid
    (b) long–chain alcohol and a sodium salt of a fatty acid
    (c) glycerol and a fatty acid
    (d) glycerol and a sodium salt of a fatty acid
    (e) none of the above

Section 20.6 *Nucleic Acids*

29. Which of the following is found in both DNA and RNA nucleotides?
    (a) deoxyribose sugar
    (b) phosphoric acid
    (c) ribose sugar
    (d) thymine
    (e) uracil

30. How many nucleotide sequences are possible for a DNA trinucleotide having an cytosine base and two guanine bases?
    (a) 1
    (b) 3
    (c) 6
    (d) 9
    (e) none of the above

31. During DNA replication, an adenine base on the template DNA strand will code for which base in the complementary DNA strand?
    (a) cytosine
    (b) guanine
    (c) thymine
    (d) uracil
    (e) the base varies

32. During RNA transcription, an adenine base on the template DNA strand will code for which base in the growing RNA strand?
    (a) cytosine
    (b) guanine
    (c) thymine
    (d) uracil
    (e) the base varies

Protein Synthesis

33. Explain how tRNA builds a protein chain using mRNA as a template.

Chapter 20 Key Terms

Across
2. an –O– bond that joins two simple sugars
5. a polymer of simple sugar units
7. a reaction of a triglyceride and sodium hydroxide
8. an ester of glycerol, 2 fatty acids, and H_3PO_4
16. a simple sugar such as glucose or fructose
18. an ester of a long-chain alcohol and a fatty acid
21. a lipid that contains mostly unsaturated fatty acids
22. a compound that catalyzes a biochemical reaction
23. a polymer of amino acids
24. an acid that carries genetic information

Down
1. a type of bond that joins two amino acids
3. a lipid containing four rings of carbon atoms
4. a lipid that contains mostly saturated fatty acids
6. a water-insoluble compound such as a fat or an oil
9. two amino acids joined by an amide bond
10. the study of biological compounds and reactions
11. an ester of glycerol and three fatty acids
12. 50 amino acids linked in a long-chain molecule
13. an acid with a long hydrocarbon side-chain
14. a sugar that hydrolyzes to give two sugars
15. a polyhydroxy aldehyde or ketone
17. a carboxylic acid containing an $-NH_2$ group
19. a repeating unit in a nucleic acid
20. a large compound with repeating chemical units

Chapter 20 Answers to Self–Test

1. **a** 2. **b** 3. **c** 4. **d** 5. **a** 6. **b** 7. **c** 8. **d** 9. **c** 10. **b** 11. **e** 12. **c** 13. **b** 14. **c** 15. **c** 16. **a**
17. **c** 18. **b** 19. **b** 20. **b** 21. **b** 22. **a** 23. **c** 24. **d** 25. **d** 26. **c** 27. **d** 28. **b** 29. **b** 30. **b**
31. **c** 32. **d**

Protein Synthesis

33. First, a molecule of mRNA is made in the cell nucleus using a section of DNA.
    A portion of the DNA molecule acts as a template to generate a sequence of
    organic bases in the molecule of mRNA. Second, the mRNA molecule moves
    out of the nucleus and gives instructions to transfer RNA (tRNA) for the amino
    acid sequence in a particular protein chain.

Chapter 20 Answers to Crossword Puzzle

Across
- 2. glycoside
- 5. polysaccharide
- 7. saponification
- 8. phospholipid
- 16. monosaccharide
- 18. wax
- 21. oil
- 22. enzyme
- 23. protein
- 24. nucleic

Down
- 1. peptide
- 3. steroid
- 4. fat
- 6. lipid
- 9. dipeptide
- 10. biochemistry
- 11. triglyceride
- 12. polypeptide
- 13. fatty
- 14. disaccharide
- 15. carbohydrate
- 17. amino
- 19. nucleotide
- 20. polymer

Solutions Manual

to

Odd–Numbered Exercises

Introduction to Chemistry

CHAPTER

1

Section 1.1 *Evolution of Chemistry*

1. yin and yang

3. Robert Boyle

5. A *hypothesis* is an initial proposal that is tentative, whereas a *theory* is a proposal that has been extensively tested.

7. Statement (a) is a *theory* because it explains the composition of an atom.
   Statement (b) is a *theory* because it explains a change in the neutron.
   Statement (c) is a *natural law* because gas pressures can be measured.
   Statement (d) is a *natural law* because gas volumes can be measured.

Section 1.2 *Modern Chemistry*

9. Antoine Lavoisier

11. Doctors and nurses receive training in chemistry as well as dentists, physical therapists, chiropractors, dietitians, veterinarians, scientists, engineers, and many others.

13. A solution to the nine–dot problem using three straight lines is shown below; the assumption regards the angle of the lines and the size of the dots.

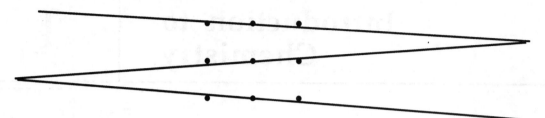

15. By flipping the image, we can view the blocks as stacking upward or as hanging downward.

CHAPTER

Scientific Measurements

2

Section 2.1 *Uncertainty in Measurements*

1.
	Instrument	Quantity		Instrument	Quantity
(a)	metric ruler	length	(b)	buret	volume
(c)	balance	mass	(d)	pipet	volume
(e)	stopwatch	time	(f)	graduated cylinder	volume

3.
	Maximum	Minimum		Maximum	Minimum
(a)	6.6 cm	6.4 cm	(b)	0.52 g	0.50 g
(c)	10.1 mL	9.9 mL	(d)	35.6 s	35.4 s

Section 2.2 *Significant Digits*

5.
	Measurement	Significant Digits
(a)	0.05 cm	1 significant digit
(b)	0.707 g	3 significant digits
(c)	83.0 mL	3 significant digits
(d)	34.60 s	4 significant digits

7.
	Measurement	Significant Digits
(a)	280 cm	2 significant digits
(b)	1200 g	2 significant digits
(c)	100 mL	1 significant digit
(d)	3000 s	1 significant digit

9.
	Measurement	Significant Digits
(a)	3.71×10^3 cm	3 significant digits
(b)	9.5×10^{-1} g	2 significant digits
(c)	2.000×10^4 mL	4 significant digits
(d)	1×10^{-9} s	1 significant digit

11.

	Example	Significant Digits
(a)	25 metric rulers	0 (not a measurement)
(b)	2 metric balances	0 (not a measurement)
(c)	12 beakers	0 (not a measurement)
(d)	10 stopwatches	0 (not a measurement)

Section 2.3 *Rounding Off Nonsignificant Digits*

13.

	Example	Rounded Off
(a)	31.505	31.5
(b)	213,600	214,000
(c)	5.155	5.16
(d)	77.504	77.5

15.

	Example	Rounded Off
(a)	61.15	61.2
(b)	362.01	362
(c)	2155	2160
(d)	0.3665	0.367

Section 2.4 *Adding and Subtracting Measurements*

17.
- (a) $31.15 \text{ cm} + 41.000 \text{ cm} = 72.150 \text{ cm}$ *rounds to* 72.15 cm
- (b) $50.2 \text{ cm} - 0.500 \text{ cm} = 49.700 \text{ cm}$ *rounds to* 49.7 cm

19.
- (a) $0.4 \text{ g} + 0.44 \text{ g} + 0.444 \text{ g} = 1.284 \text{ g}$ *rounds to* 1.3 g
- (b) $15.5 \text{ g} + 7.50 \text{ g} + 0.050 \text{ g} = 23.050 \text{ g}$ *rounds to* 23.1 g

21.
- (a) $242.167 \text{ g} - 175 \text{ g} = 67.167 \text{ g}$ *rounds to* 67 g
- (b) $27.55 \text{ g} - 14.545 \text{ g} = 13.005 \text{ g}$ *rounds to* 13.01 g

Section 2.5 *Multiplying and Dividing Measurements*

23.
- (a) $3.65 \text{ cm} \times 2.10 \text{ cm} = 7.665 \text{ cm}^2$ *rounds to* 7.67 cm^2
- (b) $8.75 \text{ cm} \times 1.15 \text{ cm} = 10.0625 \text{ cm}^2$ *rounds to* 10.1 cm^2
- (c) $16.5 \text{ cm} \times 1.7 \text{ cm} = 28.05 \text{ cm}^2$ *rounds to* 28 cm^2
- (d) $21.1 \text{ cm} \times 20 \text{ cm} = 422 \text{ cm}^2$ *rounds to* 400 cm^2

25.
- (a) $\dfrac{66.3 \text{ g}}{7.5 \text{ mL}} = 8.84 \text{ g/mL}$ *rounds to* 8.8 g/mL
- (b) $\dfrac{12.5 \text{ g}}{4.1 \text{ mL}} = 3.04878 \text{ g/mL}$ *rounds to* 3.0 g/mL
- (c) $\dfrac{42.620 \text{ g}}{10.0 \text{ mL}} = 4.262 \text{ g/mL}$ *rounds to* 4.26 g/mL
- (d) $\dfrac{91.235 \text{ g}}{10.00 \text{ mL}} = 9.1235 \text{ g/mL}$ *rounds to* 9.124 g/mL

Section 2.6 *Exponential Numbers*

27. (a) $2 \times 2 \times 2 \times 2 \times 2 \times 2 = 2^6$

    (b) $\frac{1}{2} \times \frac{1}{2} \times \frac{1}{2} = (\frac{1}{2})^3 = 2^{-3}$

29.

    (a) $10 \times 10 \times 10 \times 10 \times 10 = 10^5$

    (b) $\frac{1}{10} \times \frac{1}{10} \times \frac{1}{10} = (\frac{1}{10})^3 = 10^{-3}$

31.

	Scientific Notation	Ordinary Number
(a)	1×10^3	1000
(b)	1×10^{-7}	0.000 000 1

33.

	Ordinary Number	Scientific Notation
(a)	1,000,000,000	1×10^9
(b)	0.000 000 01	1×10^{-8}

35.

	Scientific Notation	Ordinary Number
(a)	1×10^1	10
(b)	1×10^{-1}	0.1

Section 2.7 *Scientific Notation*

37.

	Ordinary Number	Scientific Notation
(a)	80,916,000	8.0916×10^7
(b)	0.000 000 015	1.5×10^{-8}
(c)	335,600,000,000,000	3.356×10^{14}
(d)	0.000 000 000 000 927	9.27×10^{-13}

39. 2.69×10^{22} neon atoms

Section 2.8 *Unit Equations and Unit Factors*

41.

	Unit Equation
(a)	2 nickels = 1 dime
(b)	1 nickel = 5 pennies

43.

	Unit Equation
(a)	1 mile = 5280 feet
(b)	2000 pounds = 1 ton

45. <u>Unit Factors</u>

(a) $\dfrac{2 \text{ nickels}}{1 \text{ dime}}$ and $\dfrac{1 \text{ dime}}{2 \text{ nickels}}$

(b) $\dfrac{1 \text{ nickel}}{5 \text{ pennies}}$ and $\dfrac{5 \text{ pennies}}{1 \text{ nickel}}$

47. (a) $\dfrac{1 \text{ mile}}{5280 \text{ feet}}$ and $\dfrac{5280 \text{ feet}}{1 \text{ mile}}$

(b) $\dfrac{2000 \text{ pounds}}{1 \text{ ton}}$ and $\dfrac{1 \text{ ton}}{2000 \text{ pounds}}$

49. (a) 1 mile $\equiv$ 5280 feet   (b) 1 kilogram $\cong$ 2.20 pounds
    (c) 1 liter $\cong$ 1.06 quarts   (d) 1 week $\equiv$ 7 days

51. (a) 3 significant digits   (b) infinite significant digits
    (c) infinite significant digits   (d) 2 significant digits

Section 2.9 *Unit Analysis Problem Solving*

53. $15 \cancel{\text{ miles}} \times \dfrac{8 \text{ furlongs}}{1 \cancel{\text{ mile}}} = 120 \text{ furlongs}$

55. $0.750 \cancel{\text{ carat}} \times \dfrac{0.200 \text{ gram}}{1 \cancel{\text{ carat}}} = 0.150 \text{ grams}$

57. $2.5 \cancel{\text{ pints}} \times \dfrac{2 \text{ cups}}{1 \cancel{\text{ pint}}} = 5.0 \text{ cups}$

59. $3.25 \cancel{\text{ years}} \times \dfrac{12 \text{ months}}{1 \cancel{\text{ year}}} = 39.0 \text{ months}$

61. $263{,}000{,}000{,}000 \cancel{\text{ troy ounces}} \times \dfrac{1 \text{ troy pounds}}{12 \cancel{\text{ troy ounces}}} = 2.19 \times 10^{10} \text{ troy pounds}$

63. $\dfrac{186\,000 \text{ miles}}{1 \cancel{\text{ second}}} \times \dfrac{60 \cancel{\text{ seconds}}}{1 \cancel{\text{ minute}}} \times \dfrac{60 \cancel{\text{ minutes}}}{1 \cancel{\text{ hour}}} \times \dfrac{24 \cancel{\text{ hours}}}{1 \cancel{\text{ day}}} \times \dfrac{365 \cancel{\text{ days}}}{1 \text{ year}}$

$= \dfrac{5.87 \times 10^{12} \text{ miles}}{\text{year}}$

Section 2.10 *The Percent Concept*

65. $\dfrac{101 \text{ \sout{students}}}{5846 \text{ \sout{students}}} \times 100 = 1.73\%$ are chemistry majors

Note: This is not a measurement and significant digits do not apply.

67. $5.750 \text{ \sout{g bauxite}} \times \dfrac{4.10 \text{ g aluminum}}{100 \text{ \sout{g bauxite}}} = 0.236$ g aluminum

69. 255 mL ethanol + 375 mL water = 630 mL solution

$\dfrac{255 \text{ \sout{mL}}}{630 \text{ \sout{mL}}} \times 100 = 40.5\%$ of solution is ethanol

71. $15.0 \text{ \sout{g oxygen}} \times \dfrac{100 \text{ g water}}{88.8 \text{ \sout{g oxygen}}} = 16.9$ g water

73. Oxygen: $2.37 \times 10^{25} \text{ \sout{g crust}} \times \dfrac{49.2 \text{ g oxygen}}{100 \text{ \sout{g crust}}} = 1.17 \times 10^{25}$ g oxygen

Silicon: $2.37 \times 10^{25} \text{ \sout{g crust}} \times \dfrac{25.7 \text{ g silicon}}{100 \text{ \sout{g crust}}} = 6.09 \times 10^{24}$ g silicon

Aluminum: $2.37 \times 10^{25} \text{ \sout{g crust}} \times \dfrac{7.50 \text{ g aluminum}}{100 \text{ \sout{g crust}}} = 1.78 \times 10^{24}$ g aluminum

75. Mass of coin = 3.051 g (copper and zinc)
Mass of copper = 2.898 g
Mass of zinc = mass of coin – mass of copper
Mass of zinc = 3.051 g – 2.898 g = 0.153 g

Percent of copper: $\dfrac{2.898 \text{ \sout{g}}}{3.051 \text{ \sout{g}}} \times 100 = 94.99\%$ copper

Percent of zinc: $\dfrac{0.153 \text{ \sout{g}}}{3.051 \text{ \sout{g}}} \times 100 = 5.01\%$ zinc

General Exercises

77. 10.0 mL ($\pm$ 0.1 mL)

79. 23.0 atomic mass units

81. 126.457 g + 131.6 g = 258.057 g   *rounds to* 258.1 g

83.

Exponential Number	Scientific Notation
(a) 352×10^4	3.52×10^6
(b) 0.191×10^{-5}	1.91×10^{-6}

85. Total mass: 9.10953×10^{-28} g $+ 1.67265 \times 10^{-24}$ g $= 1.67356 \times 10^{-24}$ g

87. 1 metric ton $=$ 2200 pounds
1 English ton $=$ 2000 pounds
Mass difference: 2200 pounds $-$ 2000 pounds $=$ 200 pounds

89. 4.0 billion years $= 4.0 \times 10^9$ years

$$4.0 \times 10^9 \; \text{years} \; \times \; \frac{365 \; \text{days}}{1 \; \text{year}} \times \frac{24 \; \text{hours}}{1 \; \text{day}} = 3.5 \times 10^{13} \; \text{hours}$$

91. $$1 \; \text{pound} \; \times \; \frac{16 \; \text{ounces}}{1 \; \text{pound}} \times \frac{28.4 \; \text{grams}}{1 \; \text{ounce}} = 454 \; \text{grams of feathers}$$

93. 1 pound $=$ 454 grams of feathers
1 troy pound $=$ 373 grams of gold

Since 454 grams is greater than 373 grams, a pound of feathers weighs more than a pound of gold.

The Metric System

Section 3.1 *Basic Units and Symbols*

1. All of the given statements regarding the metric system are true.

3.
	Quantity	Metric Unit	Symbol
(a)	length	meter	m
(b)	mass	kilogram	kg
(c)	volume	liter	L
(d)	time	second	s

5.
	Quantity	Metric Prefix	Symbol
(a)	1000	kilo–	k
(b)	1,000,000	mega–	M
(c)	0.1	deci–	d
(d)	0.000 001	micro–	μ

7.
	Symbol	Metric Unit		Symbol	Metric Unit
(a)	km	kilometer	(b)	Gg	gigagram
(c)	μL	microliter	(d)	ms	millisecond

9.
	Symbol	Metric Unit		Symbol	Metric Unit
(a)	mm	millimeter	(b)	kg	kilogram
(c)	mL	milliliter	(d)	μs	microsecond

Section 3.2 *Metric Conversion Factors*

11.
	Unit Equation		Unit Equation
(a)	$1000 \text{ mm} = 1 \text{ m}$	(b)	$1 \text{ g} = 1 \times 10^6 \text{ μg}$
(c)	$1 \text{ L} = 100 \text{ cL}$	(d)	$1 \times 10^9 \text{ s} = 1 \text{ Gs}$

13.

$\underline{\text{Unit Factors}}$

(a) $\dfrac{1 \times 10^9 \text{ m}}{1 \text{ Gm}}$ and $\dfrac{1 \text{ Gm}}{1 \times 10^9 \text{ m}}$

$\underline{\text{Unit Factors}}$

(b) $\dfrac{1 \text{ g}}{1000 \text{ mg}}$ and $\dfrac{1000 \text{ mg}}{1 \text{ g}}$

(c) $\dfrac{1 \times 10^6 \text{ μL}}{1 \text{ L}}$ and $\dfrac{1 \text{ L}}{1 \times 10^6 \text{ μL}}$

(d) $\dfrac{100 \text{ cs}}{1 \text{ s}}$ and $\dfrac{1 \text{ s}}{100 \text{ cs}}$

Section 3.3 *Metric–Metric Conversions*

15. (a) $1.55 \text{ km} \times \dfrac{1000 \text{ m}}{1 \text{ km}} = 1550 \text{ m}$

(b) $0.486 \text{ g} \times \dfrac{100 \text{ cg}}{1 \text{ g}} = 48.6 \text{ cg}$

(c) $125 \text{ mL} \times \dfrac{1 \text{ L}}{1000 \text{ mL}} = 0.125 \text{ L}$

(d) $100 \text{ ns} \times \dfrac{1 \text{ s}}{1 \times 10^9 \text{ ns}} = 1 \times 10^{-7} \text{ s}$

17. (a) $125 \text{ Gm} \times \dfrac{1 \times 10^9 \text{ m}}{1 \text{ Gm}} \times \dfrac{1 \text{ Mm}}{1 \times 10^6 \text{ m}} = 1.25 \times 10^5 \text{ Mm}$

(b) $255 \text{ mg} \times \dfrac{1 \text{ g}}{1000 \text{ mg}} \times \dfrac{10 \text{ dg}}{1 \text{ g}} = 2.55 \text{ dg}$

(c) $14.5 \text{ μL} \times \dfrac{1 \text{ L}}{1 \times 10^6 \text{ μL}} \times \dfrac{100 \text{ cL}}{1 \text{ L}} = 1.45 \times 10^{-3} \text{ cL}$

(d) $1.56 \times 10^{-3} \text{ ks} \times \dfrac{1000 \text{ s}}{1 \text{ ks}} \times \dfrac{1 \times 10^9 \text{ ns}}{1 \text{ s}} = 1.56 \times 10^9 \text{ ns}$

19. $35 \text{ ms} \times \dfrac{1 \text{ s}}{1000 \text{ ms}} \times \dfrac{1 \times 10^6 \text{ μs}}{1 \text{ s}} = 3.5 \times 10^4 \text{ μs}$

Section 3.4 *Metric–English Conversions*

21. (a) $2.54 \text{ cm} = 1 \text{ in.}$ (b) $454 \text{ g} = 1 \text{ lb}$
    (c) $946 \text{ mL} = 1 \text{ qt}$ (d) $1 \text{ s} = 1 \text{ sec}$

23. (a) $66 \text{ in.} \times \dfrac{2.54 \text{ cm}}{1 \text{ in.}} = 170 \text{ cm}$

(b) $1.01 \text{ lb} \times \dfrac{454 \text{ g}}{1 \text{ lb}} = 459 \text{ g}$

(c) $0.500 \text{ qt} \times \dfrac{946 \text{ mL}}{1 \text{ qt}} = 473 \text{ mL}$

(d) $8.00 \times 10^2 \text{ sec} \times \dfrac{1 \text{ s}}{1 \text{ sec}} = 8.00 \times 10^2 \text{ s}$

25. (a) $72 \text{ in.} \times \dfrac{2.54 \text{ cm}}{1 \text{ in.}} \times \dfrac{1 \text{ m}}{100 \text{ cm}} = 1.8 \text{ m}$

(b) $175 \text{ lb} \times \dfrac{454 \text{ g}}{1 \text{ lb}} \times \dfrac{1 \text{ kg}}{1000 \text{ g}} = 79.5 \text{ kg}$

(c) $1250 \text{ mL} \times \dfrac{1 \text{ qt}}{946 \text{ mL}} \times \dfrac{1 \text{ gallon}}{4 \text{ qts}} = 0.330 \text{ gallons}$

(d) $1.52 \times 10^3 \text{ ds} \times \dfrac{1 \text{ s}}{10 \text{ ds}} \times \dfrac{1 \text{ min}}{60 \text{ s}} = 2.53 \text{ min}$

27. $\dfrac{52 \text{ mi}}{1 \text{ gal}} \times \dfrac{1 \text{ gal}}{4 \text{ qt}} \times \dfrac{1 \text{ qt}}{946 \text{ mL}} \times \dfrac{1000 \text{ mL}}{1 \text{ L}} \times \dfrac{1.61 \text{ km}}{1 \text{ mi}} = 22 \text{ km/L}$

Section 3.5 *Volume by Calculation*

29. $3.05 \text{ m} \times \dfrac{100 \text{ cm}}{1 \text{ m}} = 305 \text{ cm}$

$5.08 \text{ cm} \times 10.2 \text{ cm} \times 305 \text{ cm} = 1.58 \times 10^4 \text{ cm}^3$

31. $\dfrac{(15.3 \text{ cm}^3)}{(4.95 \text{ cm})(2.45 \text{ cm})} = 1.26 \text{ cm}$

33. (a) $1 \text{ L} \times \dfrac{1000 \text{ mL}}{1 \text{ L}} = 1000 \text{ mL}$

(b) $1 \text{ L} \times \dfrac{1000 \text{ mL}}{1 \text{ L}} \times \dfrac{1 \text{ cm}^3}{1 \text{ mL}} = 1000 \text{ cm}^3$

35. $415 \text{ in.}^3 \times \left(\dfrac{2.54 \text{ cm}}{1 \text{ in.}}\right)^3 \times \dfrac{1 \text{ mL}}{1 \text{ cm}^3} \times \dfrac{1 \text{ L}}{1000 \text{ mL}} = 6.80 \text{ L}$

Section 3.6 *Volume by Displacement*

37. volume of gemstone: $5.0 \text{ mL} - 4.5 \text{ mL} = 0.5 \text{ mL}$

39. volume of hydrogen: $255 \text{ mL} - 125 \text{ mL} = 130 \text{ mL}$

Section 3.7 *The Density Concept*

41.

	Substance	Sink or Float			Substance	Sink or Float
(a)	ebony wood	sink		(b)	bamboo	float

43.

	Gas	Rise or Fall			Gas	Rise or Fall
(a)	helium	rise		(b)	laughing gas	fall

45. (a) $250 \, \cancel{mL} \times \dfrac{0.69 \, g}{1 \, \cancel{mL}} = 170 \, g$

(b) $0.75 \, \cancel{cm^3} \times \dfrac{2.18 \, g}{1 \, \cancel{cm^3}} = 1.6 \, g$

47. (a) $0.500 \, \cancel{g} \times \dfrac{1 \, mL}{3.12 \, \cancel{g}} = 0.160 \, mL$

(b) $10.0 \, \cancel{g} \times \dfrac{1 \, \cancel{cm^3}}{8.90 \, \cancel{g}} \times \dfrac{1 \, mL}{1 \, \cancel{cm^3}} = 1.12 \, mL$

49. (a) $\dfrac{19.7 \, g}{25.0 \, mL} = 0.788 \, g/mL$

(b) $\dfrac{11.6 \, g}{4.1 \, mL} = 2.8 \, g/mL$

Section 3.8 *Temperature*

51.

	Scale	Freezing Point of Water
(a)	Fahrenheit	32°F
(b)	Kelvin	273 K
(c)	Celsius	0°C

53. (a) $(100 - 32) \, \cancel{°F} \times \dfrac{100°C}{180 \, \cancel{°F}} = 38°C$

(b) $(-215 - 32) \, \cancel{°F} \times \dfrac{100°C}{180 \, \cancel{°F}} = -137°C$

55. (a) $\left(495 \, \cancel{°C} \times \dfrac{1 \, K}{1 \, \cancel{°C}}\right) + 273 \, K = 768 \, K$

(b) $\left(-185 \, \cancel{°C}\right) \times \dfrac{1 \, K}{1 \, \cancel{°C}} + 273 \, K = 88 \, K$

Section 3.9 *Heat and Specific Heat*

57. Temperature is a measure of the average kinetic energy in a system and heat is a measure of the total energy in a system. (For example, a large chemistry lecture room at 20°C has more heat energy than a small laboratory room at 20°C, even though the temperature is the same.)

59. $250 \text{ g} \times \dfrac{1.00 \text{ cal}}{1 \text{ g} \times °C} \times (100 - 23)°C = 19,000 \text{ cal} \ (1.9 \times 10^4 \text{ cal})$

61. $25.0 \text{ g} \times \dfrac{0.108 \text{ cal}}{1 \text{ g} \times °C} \times (50.0 - 25.0)°C = 67.5 \text{ cal}$

63. $\dfrac{25.0 \text{ cal}}{30.0 \text{ g} (54.9 - 27.7)°C} = 0.0306 \text{ cal}/\text{g} \times °C$

65. $\dfrac{75.6 \text{ cal}}{(31.4 - 20.7)°C} \times \dfrac{1 \text{ g} \times °C}{0.125 \text{ cal}} = 56.5 \text{ g}$

67. $\dfrac{35.2 \text{ cal}}{10.5 \text{ g}} \times \dfrac{1 \text{ g} \times °C}{0.0566 \text{ cal}} = 59.2 \ °C$

General Exercises

69. $270 \text{ MB} \times \dfrac{1 \times 10^6 \text{ Bytes}}{1 \text{ MB}} \times \dfrac{1 \text{ kB}}{1 \times 10^3 \text{ Bytes}} = 2.7 \times 10^5 \text{ kB}$

   $2.7 \times 10^5 \text{ kB} \times \dfrac{1 \text{ disc}}{720 \text{ kB}} = 380 \text{ discs}$

71. (a) an infinite number (*exact by definition*) (b) 3 significant digits
   (c) an infinite number (*exact by definition*) (d) 3 significant digits

73. $\dfrac{186\,000 \text{ mi}}{1 \text{ s}} \times \dfrac{60 \text{ s}}{1 \text{ min}} \times \dfrac{60 \text{ min}}{1 \text{ h}} \times \dfrac{24 \text{ h}}{1 \text{ d}} \times \dfrac{365 \text{ d}}{1 \text{ yr}} = 5.87 \times 10^{12} \text{ mi}/\text{yr}$

75. $1500 \text{ m} \times \dfrac{1 \text{ km}}{1000 \text{ m}} \times \dfrac{1 \text{ mile}}{1.61 \text{ km}} = 0.93 \text{ mile}$

   The 1500–meter race is shorter than a mile, thus the time is faster.

77. $2.50 \text{ kg} \times \dfrac{1000 \text{ g}}{1 \text{ kg}} \times \dfrac{1000 \text{ mg}}{1 \text{ g}} \times \dfrac{1 \text{ grain}}{64.8 \text{ mg}} \times \dfrac{1 \text{ tablet}}{5 \text{ grains}} = 7720 \text{ tablets}$

79. $160.0 \text{ feet} \times \dfrac{1 \text{ yard}}{3 \text{ feet}} = 53.33 \text{ yard}$

   Area of football field: $53.33 \text{ yd} \times 100.0 \text{ yd} = 5333 \text{ yd}^2$

81. Volume of fool's gold: $57.5 \text{ mL} - 50.0 \text{ mL} = 7.5 \text{ mL}$

Density of fool's gold: $\dfrac{37.51 \text{ g}}{7.5 \text{ mL}} = 5.0 \text{ g/mL}$

83. Specific gravity of gasohol: $\dfrac{0.801 \text{ g/ml}}{1.00 \text{ g/ml}} = 0.801$

85. $\dfrac{1.00 \text{ g}}{1 \text{ mL}} \times \dfrac{1 \text{ lb}}{454 \text{ g}} \times \dfrac{1 \text{ mL}}{1 \text{ cm}^3} \times \left(\dfrac{2.54 \text{ cm}}{1 \text{ in.}}\right)^3 \times \left(\dfrac{12 \text{ in.}}{1 \text{ ft}}\right)^3 = 62.4 \text{ lb/ft}^3$

87. $(-422 - 32)\text{°F} \times \dfrac{100\text{°C}}{180\text{°F}} = -252\text{°C}$

$\left(-252\text{°C} \times \dfrac{1 \text{ K}}{1\text{°C}}\right) + 273 \text{ K} = 21 \text{ K}$

89. Heat gained by water = Heat lost by poker

Heat gained by water: $3500 \text{ g} \times \dfrac{1.00 \text{ cal}}{1 \text{ g} \times \text{°C}} \times (51 - 22)\text{°C} = 1.0 \times 10^5 \text{ cal}$

$1.0 \times 10^5 \text{ cal} \times \dfrac{1 \text{ kcal}}{1000 \text{ cal}} = 1.0 \times 10^2 \text{ kcal of heat are lost by the poker}$

91. $10.0 \text{ g} \times \dfrac{0.0920 \text{ cal}}{1 \text{ g} \cdot \text{°C}} \times (T_{final} - 22.7)\text{°C} = 27.8 \text{ cal}$

$T_{final} = \left(\dfrac{27.8 \text{ cal}}{10.0 \text{ g}} \times \dfrac{1 \text{ g} \times \text{°C}}{0.0920 \text{ cal}}\right) + 22.7\text{°C} = 52.9\text{°C}$

93. $\dfrac{13.6 \text{ g}}{1 \text{ mL}} \times \dfrac{1 \text{ kg}}{1000 \text{ g}} \times \dfrac{1 \text{ mL}}{1 \text{ cm}^3} \times \left(\dfrac{100 \text{ cm}}{1 \text{ m}}\right)^3 = 1.36 \times 10^4 \text{ kg/m}^3$

95. Volume $= \pi r^2 h$
    $r = 1.95 \text{ cm}$

Volume $= 1 \text{ kg} \times \dfrac{1000 \text{ g}}{1 \text{ kg}} \times \dfrac{\text{cm}^3}{21.50 \text{ g}} = 46.51 \text{ cm}^3$

$\pi r^2 h = 46.51 \text{ cm}^3$

$h = \dfrac{46.51 \text{ cm}^3}{\pi \cdot r^2} = \dfrac{46.51 \text{ cm}^3}{3.14 \times (1.95 \text{ cm})^2} = 3.90 \text{ cm}$

Note: diameter $= 2 \times$ radius $= 2 \times 1.95 \text{ cm} = 3.90 \text{ cm}$
Thus, the diameter and the height of the International Prototype Kilogram are equal.

Matter and Energy

Section 4.1 *Physical States of Matter*

1.
State	Shape
(a) solids	definite
(b) liquids	indefinite
(c) gases	indefinite

3.
State	Compressibility
(a) solid	negligible
(b) liquid	negligible
(c) gas	significant

5.
Change in State	Term
(a) solid to liquid	melting
(b) liquid to gas	vaporizing
(c) gas to solid	deposition

7.
Change	Heat Energy
(a) solid to liquid	absorbed
(b) liquid to gas	absorbed
(c) gas to solid	released

Section 4.2 *Elements, Compounds, and Mixtures*

9. Homogeneous mixtures have definite properties and variable composition, while heterogeneous mixtures have indefinite properties and variable composition.

   homogeneous mixture: 5¢ nickel (coin is an alloy of nickel and zinc)
   heterogeneous mixture: 1¢ penny (new coin is zinc plated with copper)

11.

Substance	Classification
(a) Earth's crust	heterogeneous mixture
(b) Earth's atmosphere	homogenous mixture
(c) Earth's oceans	homogenous mixture
(d) Earth's rivers	homogenous mixture

Note: The Earth's oceans and rivers can be considered heterogeneous mixtures if they contain solid material such as ice or seaweed.

13.

Substance	Classification
(a) iron metal	element
(b) iron ore	mixture
(c) iron oxide	compound
(d) steel alloy	mixture

Section 4.3 *Names and Symbols of the Elements*

15. The three most abundant elements in the Earth's crust are oxygen, silicon, and aluminum.

17.

Element	Symbol		Element	Symbol
(a) lithium	Li	(b)	argon	Ar
(c) magnesium	Mg	(d)	manganese	Mn
(e) fluorine	F	(f)	sodium	Na
(g) copper	Cu	(h)	nickel	Ni

19.

Symbol	Element		Symbol	Element
(a) Cl	chlorine	(b)	Ne	neon
(c) Cd	cadmium	(d)	Ge	germanium
(e) Co	cobalt	(f)	Ra	radium
(g) Cr	chromium	(h)	Te	tellurium

21.

Element	Atomic no.		Element	Atomic no.
(a) hydrogen	1	(b)	barium	56
(c) gold	79	(d)	iodine	53
(e) bromine	35	(f)	aluminum	13
(g) potassium	19	(h)	tin	50

Section 4.4 *Metals, Nonmetals, and Semimetals*

23.

Property	Classification
(a) shiny luster	metal
(b) low melting point	nonmetal
(c) malleable	metal
(d) reacts with metals and nonmetals	nonmetal

25.

	Property	Classification
(a)	shiny solid	metal
(b)	brittle solid	nonmetal
(c)	ductile	metal
(d)	reacts with calcium	nonmetal

27.

	Element	Classification
(a)	Na	metal
(b)	Br	nonmetal
(c)	Ar	nonmetal
(d)	Ge	semimetal

29.

	Element	Classification
(a)	beryllium	metal
(b)	germanium	semimetal
(c)	phosphorus	nonmetal
(d)	manganese	metal

31.

	Element	Physical State
(a)	Na	solid
(b)	N	gas
(c)	Ar	gas
(d)	Ge	solid

33.

	Element	Physical State
(a)	cobalt	solid
(b)	bromine	liquid
(c)	helium	gas
(d)	titanium	solid

Section 4.5 *Compounds and Chemical Formulas*

35. From the law of constant composition, we can predict the mass ratio of the elements in a copper carbonate mineral is $5:1:4$.

37.

	Chemical Formula	Chemical Composition
(a)	$C_9H_8O_4$	9 carbon atoms 8 hydrogen atoms 4 oxygen atoms
(b)	$C_{17}H_{20}N_4O_6$	17 carbon atoms 20 hydrogen atoms 4 nitrogen atoms 6 oxygen atoms

39.   Atomic Composition   Formula
     (a)   20 carbon, 30 hydrogen, 1 oxygen $C_{20}H_{30}O$
     (b)   12 carbon, 18 hydrogen, 2 chlorine, $C_{12}H_{18}Cl_2N_4OS$
         4 nitrogen, 1 oxygen and 1 sulfur atom

41.   Chemical Formula   Total Number of Atoms
     (a) $C_3H_5OH(COOH)_3$ 22 atoms
     (b) $C_3H_4(OH)COOH$ 13 atoms

Section 4.6 *Physical and Chemical Properties*

43.   Property   Property
     (a)   physical   (b)   physical
     (c)   physical   (d)   chemical

45.   Property   Property
     (a)   physical   (b)   chemical
     (c)   chemical   (d)   physical

47.   Property   Property
     (a)   chemical   (b)   physical
     (c)   physical   (d)   physical

49.   Property   Property
     (a)   physical   (b)   chemical
     (c)   physical   (d)   chemical

Section 4.7 *Physical and Chemical Changes*

51.   Change   Change
     (a)   physical   (b)   chemical
     (c)   chemical   (d)   physical

53.   Change   Change
     (a)   physical   (b)   chemical
     (c)   physical   (d)   chemical

55.   Change   Change
     (a)   physical   (b)   chemical
     (c)   physical   (d)   chemical

57.   Change   Change
     (a)   chemical   (b)   physical
     (c)   chemical   (d)   physical

Section 4.8 *Conservation of Mass*

59. 2.50 g iron + 1.44 g sulfur = 3.94 g of iron sulfide

61. 0.750 g mercury oxide – 0.695 g mercury = 0.055 g oxygen

Section 4.9 *Potential and Kinetic Energy*

63. On a roller coaster ride the potential energy is greatest when the roller coaster is at its highest point. As the roller coaster descends, it loses potential energy and gains kinetic energy. Thus, the *potential energy is maximum*, and the *kinetic energy is minimum*, when the roller coaster is at its highest point.

65. A substance in the gaseous state has more kinetic energy than in the liquid state which has more kinetic energy than the solid state.

67. If the temperature increases, the kinetic energy will *increase*.

69. If a steel cylinder of argon gas is heated from 50° to 100° C, the kinetic energy and motion of argon atoms *increase*.

Section 4.10 *Conservation of Energy*

71. From the law of conservation of energy, 1500 kilocalories of heat energy is required to vaporize the water, at 100°C, to steam.

73. 697 calories + 110 calories = 807 calories released

75. From the law of conservation of energy, 48.0 kilocalories of heat is required to decompose 15.0 g of water into hydrogen and oxygen gases.

77.   <u>Forms of Energy</u>
    (1) heat energy
    (2) light energy
    (3) chemical energy
    (4) electrical energy
    (5) mechanical energy
    (6) nuclear energy

79.   <u>Forms of Energy</u>
    (a) nuclear → heat
    (b) heat → mechanical
    (c) mechanical → mechanical
    (d) mechanical → electrical

81.   Forms of Energy
(a)   chemical → electrical
(b)   electrical → mechanical
(c)   mechanical → mechanical
(d)   chemical → heat

83. $E = mc^2$; E = energy, m = mass, c = the speed of light (3.00×10^8 m/s)

General Exercises

85.   Gasoline is a solution of different compounds and therefore gasoline is a *homogenous* mixture.

87.   The element germanium is below silicon in Group IV/14; thus *germanium* has similar chemical properties and can be substituted for silicon.

89.   Element   Chemical Symbol
(a)   ferrum   Fe
(b)   plumbum   Pb
(c)   stannum   Sn
(d)   aurum   Au

91.   A change of physical state but not chemical formula is an example of a *physical change*.

93.   From the conservation of energy law, 3250 calories of heat is required to decompose 5.00 g of ammonia into hydrogen gas and nitrogen gas.

95.   From the conservation of mass and energy law, the products should weigh very slightly less than the reactants. In practice, the mass loss is undetectable although, in theory, a small amount of mass is converted to heat energy.

Models of the Atom

Section 5.1 *Dalton Model of the Atom*

1.   (1)   An element is composed of tiny, indivisible, indestructible particles
            called atoms.
      (2)   All atoms of an element are identical and have the same properties.
      (3)   Atoms of different elements combine to form compounds.
      (4)   Compounds contain atoms in small, whole-number ratios.
      (5)   Atoms may combine in more than one ratio to form different
            compounds.

3.   (1)   Atoms are divisible (subatomic particles have been discovered).
      (2)   All atoms of an element are not identical because elements may have
            two or more isotopes.

Section 5.2 *Thomson Model of the Atom*

5.   The simplest particle observed in cathode rays was the electron (e^-).

7.   Relative Charge of Electron   Relative Charge of Proton
                1–   1+

9.   Raisins in the plum–pudding model were analogous to electrons in atoms.

Section 5.3 *Rutherford Model of the Atom*

11.   Rutherford concluded that atoms could not be "plum–pudding" spheres as
      Thomson had previously suggested. Rutherford proposed that the mass of an
      atom exists only in a tiny fraction of its total volume. Furthermore, the mass
      is located at the center of the atom (nucleus) and has a positive charge.

13. In the Rutherford model of the atom, protons and neutrons are located in a small, dense, positively charged center, while electrons surround the nucleus.

15.
Particle	Relative Charge
electron (e^-)	$1-$
proton (p^+)	$1+$
neutron ($n°$)	0

Section 5.4 *Atomic Notation*

17.
	Isotope	Neutrons		Isotope	Neutrons
(a)	^{4}He	$4 - 2 = 2\ n°$	(b)	^{32}S	$32 - 16 = 16\ n°$
(c)	^{10}B	$10 - 5 = 5\ n°$	(d)	^{44}Ca	$44 - 20 = 24\ n°$

19.
	Isotope	Neutrons		Isotope	Neutrons
(a)	Li–7	$7 - 3 = 4\ n°$	(b)	K–40	$40 - 19 = 21\ n°$
(c)	Sr–88	$88 - 38 = 50\ n°$	(d)	Pt–195	$195 - 78 = 117\ n°$

21.

Atomic Notation	Atomic Number	Mass Number	Number of Protons	Number of Neutrons	Number of Electrons
$^{4}_{2}$He	2	4	2	2	2
$^{21}_{10}$Ne	10	21	10	11	10
$^{50}_{22}$Ti	22	50	22	28	22
$^{197}_{79}$Au	79	197	79	118	79

23.

(a) (4 $n°$ / 3 p^+) 3 e–   (b) (7 $n°$ / 6 p^+) 6 e–

(c) (8 $n°$ / 8 p^+) 8 e–   (d) (10 $n°$ / 10 p^+) 10 e–

Section 5.5 *Atomic Mass*

25. Reference isotope is carbon–12.

27. Atomic masses are expressed on a relative atomic mass scale because atoms are too small to weigh directly and their very small absolute mass values would be awkward to work with in units of grams.

29. $^{27}_{13}Al$ has only one naturally occurring isotope with the mass of 26.98 amu.

31. (a) Simple Average Mass: $\dfrac{5.0\ g + 2.0\ g}{2} = 3.5\ g$

    (b) Weighted Average Mass: $5.0\ g \left(\dfrac{100\ large\ marbles}{300\ total\ marbles}\right) + 2.0\ g \left(\dfrac{200\ small\ marbles}{300\ total\ marbles}\right)$

    $$= 1.7\ g + 1.3\ g = 3.0\ g$$

33. Li–6:   6.015 amu × 0.0742   =   0.446 amu
    Li–7:   7.016 amu × 0.9258   =   6.495 amu
                               Atomic Mass   =   6.941 amu

35. Fe–54:   53.940 amu × 0.0582   =   3.14 amu
    Fe–56:   55.935 amu × 0.9166   =   51.27 amu
    Fe–57:   56.935 amu × 0.0219   =   1.25 amu
    Fe–58:   57.933 amu × 0.0033   =   0.19 amu
                               Atomic Mass   =   55.85 amu

37. Since the atomic weight of chlorine is 35.45 amu which is closer to the mass of Cl–35, the isotope Cl–35 must have a greater abundance than Cl–37.

Section 5.6 *The Wave Nature of Light*

39. Violet light is more energetic than green or orange light.

41. Orange light has a longer wavelength than violet or green light.

43. A wavelength of 450 nm is more energetic than light with a wavelength of 550 nm or 650 nm. (A shorter wavelength corresponds to higher energy.)

45. A wavelength of 450 nm has a higher frequency than light with a wavelength of 550 nm or 650 nm. (A shorter wavelength of light has a higher frequency.)

47. The quantum particle in light energy is a *photon*.

49.
	Example	Spectrum
(a)	rainbow	continuous
(b)	line spectrum	quantized

51.
	Example	Spectrum
(a)	metric ruler	continuous
(b)	digital laser	quantized

Section 5.8 *Bohr Model of the Atom*

53. The Bohr Model of the Atom

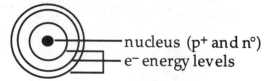

nucleus (p^+ and $n^°$)
e^- energy levels

55. The transition from energy level 5 to 2 is the most energetic because the electron drops the greatest distance.

57.
Energy Level Change	Color
4 to 2	blue–green line

59.
Radiant (Light) Energy	Energy Level Change
ultraviolet photon	5 to 1

61.
Color	Energy Level Change
red	3 to 2
blue–green	4 to 2
violet	5 to 2

The violet line is the most energetic because a violet photon corresponds to the electron dropping the greatest distance (5 to 2).

63.
	Energy Level Change	Number of Photons
(a)	3 to 1	1
(b)	3 to 2	1

65.
	Energy Level Change	Color
(a)	2 to 1	ultraviolet
(b)	3 to 2	red
(c)	4 to 3	infrared

Section 5.9 *Principal Energy Levels and Sublevels*

67. The lines in the emission spectrum of hydrogen suggest the existence of energy levels.

69.

	Principal Energy Level	Sublevels
(a)	first	$1s$
(b)	second	$2s\ 2p$
(c)	third	$3s\ 3p\ 3d$
(d)	fourth	$4s\ 4p\ 4d\ 4f$

71.

	Sublevel	Max. Electrons		Sublevel	Max. Electrons
(a)	$2s$	$2\ e^-$	(b)	$4p$	$6\ e^-$
(c)	$3d$	$10\ e^-$	(d)	$5f$	$14\ e^-$

73. The maximum number of electrons in the second energy level is equal to the sum of the maximum number of electrons in the $2s$ and $2p$ sublevels, that is, $2\ e^- + 6\ e^- = 8\ e^-$.

Section 5.10 *Electron Configuration*

75. Order of energy levels: $1s\ 2s\ 2p\ 3s\ 3p\ 4s\ 3d\ 4p\ 5s\ 4d\ 5p$

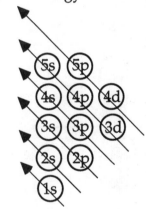

77.

	Element	Electron Configuration
(a)	He	$1s^2$
(b)	Be	$1s^2\ 2s^2$
(c)	Co	$1s^2\ 2s^2\ 2p^6\ 3s^2\ 3p^6\ 4s^2\ 3d^7$
(d)	Cd	$1s^2\ 2s^2\ 2p^6\ 3s^2\ 3p^6\ 4s^2\ 3d^{10}\ 4p^6\ 5s^2\ 4d^{10}$

79.

	Electron Configuration	Element
(a)	$1s^2\ 2s^1$	Li
(b)	$1s^2\ 2s^2\ 2p^6\ 3s^2\ 3p^2$	Si
(c)	$1s^2\ 2s^2\ 2p^6\ 3s^2\ 3p^6\ 4s^2\ 3d^2$	Ti
(d)	$1s^2\ 2s^2\ 2p^6\ 3s^2\ 3p^6\ 4s^2\ 3d^{10}\ 4p^6\ 5s^2$	Sr

81. An orbit is the path traveled by an electron of given energy about the nucleus of an atom, according to the Bohr model.

An orbital is a region about the nucleus in which there is a high probability of finding an electron of given energy, according to the quantum mechanical model.

83. (a)   (b)

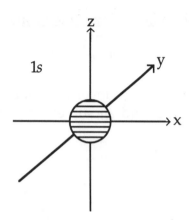
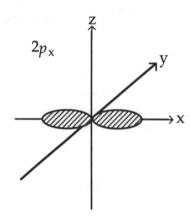

(c)   (d)

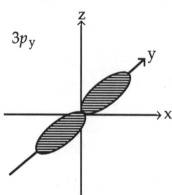
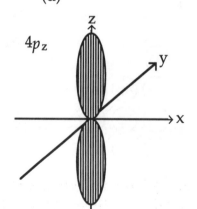

85.

	Orbitals	Higher Energy		Orbitals	Higher Energy
(a)	$2s$ or $3s$	$3s$	(b)	$2p_x$ or $3p_x$	$3p_x$
(c)	$2p_x$ or $2p_y$	both are equal	(d)	$4p_y$ or $4p_z$	both are equal

87.

	Description	Orbital
(a)	spherical orbital in the fifth shell	$5s$
(b)	dumbbell–shaped orbital in the fourth shell	$4p$

89.

Orbital	Max. # of Electrons
(a) $1s$ orbital	$2\ e^-$
(b) $2p$ orbital	$2\ e^-$
(c) $3d$ orbital	$2\ e^-$
(d) $4f$ orbital	$2\ e^-$

General Exercises

91. 1.60×10^{-19} ~~coulomb~~ $\times \dfrac{1\,\text{g}}{1.76 \times 10^8\ \text{~~coulomb~~}} = 9.09 \times 10^{-28}\,\text{g}$

93. Let X = the unknown isotopic mass of Ga–71
Percent abundance of Ga–71: $100\% - 60.10\% = 39.90\%$

Ga–69: $68.926\ \text{amu} \times 0.6010 = 41.42\ \text{amu}$

Ga–71: $X \times 0.3990 = 0.3990\ X\ \text{amu}$

Atomic Mass $= 69.72\ \text{amu}$

$41.42\ \text{amu} + 0.3990\ X = 69.72\ \text{amu}$

$0.3990\ X = 69.72\ \text{amu} - 41.42\ \text{amu}$

$0.3990\ X = 28.30\ \text{amu}$

$X = \dfrac{28.30\ \text{amu}}{0.3990} = 70.93\ \text{amu}$

The isotopic mass of Ga–71 = 70.93 amu

95. Since the periodic table lists a mass number (61), and not an atomic mass, we can conclude that there is no stable isotope of promethium.

97.

Wavelength	Region
(a) 200 nm	ultraviolet
(b) 500 nm	visible
(c) 1200 nm	infrared

99. An atom is more stable with a completely filled d sublevel than a partially filled sublevel. If one of the copper $4s$ electrons moves into the $3d$ sublevel, the $3d$ sublevel is completely filled, and thus more stable.

The Periodic Table

Section 6.1 *Classification of Elements*

1. Between chlorine and iodine: bromine (Br)
   Between calcium and barium: strontium (Sr)

3. Following the potassium series: copper (Cu)
   Following the rubidium series: silver (Ag)

Section 6.2 *The Periodic Law Concept*

5. Mendeleev suggested that the elements should be arranged according to increasing atomic mass.

7. Based on the periodic law, the chemical and physical properties of the elements tend to recur periodically when the elements are arranged according to increasing atomic number.

Section 6.3 *Groups and Periods of Elements*

9. Vertical columns in the periodic table are referred to as groups and families.

11. The main group elements, Groups IA through VIIIA, are called the representative elements.

13. The two series of elements that include Ce–Lu and Th–Lr are known as the inner transition elements.

15. The elements on the right side of the periodic table are called the nonmetals.

17.

	Family	Group Number
(a)	alkali metals	IA/1
(b)	alkaline earth metals	IIA/2
(c)	halogens	VIIA/17
(d)	noble gases	VIIIA/18

19. The elements in the series that follow element 57 are called the lanthanide series of elements.

21. The elements in Group IIIB/3 (Sc, Y, La, Ce–Lu) are called the rare earth elements.

23.

	American	IUPAC		American	IUPAC
(a)	Group IA	1	(b) Group IB	11
(c)	Group IIIA	13	(d) Group IIIB	3
(e)	Group VA	15	(f) Group VB	5
(g)	Group VIIA	17	(h) Group VIIB	7

25.

	Element		Element
(a)	Ge	(b) Na
(c)	I	(d) Pm

27.

	Element		Element
(a)	Sb	(b) Mg
(c)	Br	(d) Th

Section 6.4 *Periodic Trends*

29. Proceeding down a group of elements in the periodic table, the atomic radius of an element generally *increases*.

31. Proceeding down a group of elements in the periodic table, the metallic character of an element generally *increases*.

33.

	Atomic Radius		Atomic Radius
(a)	Li < **Na**	(b) N < **P**
(c)	Mg < **Ca**	(d) Ar < **Kr**

(*Note*: The element with the larger atomic radius is in **bold**.)

35.

	Metallic Character		Metallic Character
(a)	B < **Al**	(b) Na < **K**
(c)	Mg < **Ba**	(d) H < **Fe**

(*Note*: The element with the greater metallic character is in **bold**.)

Section 6.5 *Properties of Elements*

37. Predicted atomic radius of K: $\quad\quad$ $0.248 - (0.266 - 0.248) = \sim 0.230$ nm

Predicted density of Rb: $\quad\quad$ $\dfrac{0.86 + 1.90}{2} = \sim 1.38$ g/mL

Predicted melting point of Cs: $\quad\quad$ $38.9 - (63.3 - 38.9) = \sim 14.5°C$

39. Predicted atomic radius of Cl: $\quad\quad$ $0.115 - (0.133 - 0.115) = \sim 0.097$ nm

Predicted density of Br: $\quad\quad$ $\dfrac{1.56 + 4.97}{2} = \sim 3.27$ g/mL

Predicted boiling point of I: $\quad\quad$ $58.8 + [58.8 - (-34.6)] = \sim 152.2°C$

41.

	Compound	Formula		Compound	Formula
(a)	lithium oxide	Li_2O	(b)	calcium oxide	CaO
(c)	gallium oxide	Ga_2O_3	(d)	tin oxide	SnO_2

43.

	Compound	Formula		Compound	Formula
(a)	cadmium oxide	CdO	(b)	zinc sulfide	ZnS
(c)	mercury sulfide	HgS	(d)	cadmium selenide	$CdSe$

45.

	Compound	Formula		Compound	Formula
(a)	sulfur oxide	SO_3	(b)	tellurium oxide	TeO_3
(c)	selenium sulfide	SeS_3	(d)	tellurium sulfide	TeS_3

Section 6.6 *Blocks of Elements*

47. The elements in Groups IA/1 and IIA/2 are filling s sublevels.

49. The elements in Groups IIIB/3 through IIB/12 are filling d sublevels.

51. The elements in the lanthanide series are filling a $4f$ sublevel.

53.

	Element	Highest Sublevel		Element	Highest Sublevel
(a)	H	$1s$	(b)	Na	$3s$
(c)	Sm	$4f$	(d)	Br	$4p$
(e)	Sr	$5s$	(f)	C	$2p$
(g)	Sn	$5p$	(h)	Cs	$6s$

55.

	Element	Electron Configuration	Core Notation
(a)	Li	$1s^2\,2s^1$	[He] $2s^1$
(b)	F	$1s^2\,2s^2\,2p^5$	[He] $2s^2\,2p^5$
(c)	Mg	$1s^2\,2s^2\,2p^6\,3s^2$	[Ne] $3s^2$
(d)	P	$1s^2\,2s^2\,2p^6\,3s^2\,3p^3$	[Ne] $3s^2\,3p^3$
(e)	Ca	$1s^2\,2s^2\,2p^6\,3s^2\,3p^6\,4s^2$	[Ar] $4s^2$
(f)	Mn	$1s^2\,2s^2\,2p^6\,3s^2\,3p^6\,4s^2\,3d^5$	[Ar] $4s^2\,3d^5$
(g)	Ga	$1s^2\,2s^2\,2p^6\,3s^2\,3p^6\,4s^2\,3d^{10}\,4p^1$	[Ar] $4s^2\,3d^{10}\,4p^1$
(h)	Rb	$1s^2\,2s^2\,2p^6\,3s^2\,3p^6\,4s^2\,3d^{10}\,4p^6\,5s^1$	[Kr] $5s^1$

Section 6.7 *Valence Electrons*

57.

	Group	Valence Electrons		Group	Valence Electrons
(a)	IA/1	1	(b)	IIIA/13	3
(c)	VA/15	5	(d)	VIIA/17	7

59.

	Element	Valence Electrons		Element	Valence Electrons
(a)	H	1	(b)	B	3
(c)	N	5	(d)	F	7
(e)	Ca	2	(f)	Si	4
(g)	O	6	(h)	Ar	8

Section 6.8 *Electron Dot Formulas*

61.

	Element	Electron Dot Formula		Element	Electron Dot Formula
(a)	H	H·	(b)	B	·Ḅ·
(c)	N	·N̤:	(d)	F	:F̤:
(e)	Ca	C̤a·	(f)	Si	·S̤i·
(g)	O	·Ö:	(h)	Ar	:A̤r:

Section 6.9 *Ionization Energy*

63. Proceeding down a group of elements in the periodic table, the first ionization energy generally *decreases*.

65. The Group VIIIA/18 elements have the *highest* ionization energy.

67.   Ionization Energy

(a) **Mg** > Ca   (b) **S** > Se
(c) **Sn** > Pb   (d) **N** > P

(*Note*: The element with the higher ionization energy is in **bold**.)

69.   Ionization Energy

(a) Rb > **Cs**   (b) He > **Ar**
(c) B > **Al**   (d) F > **I**

(*Note*: The element with the lower ionization energy is in **bold**.)

Section 6.10 *Ionic Charges*

71.   Group   Ionic Charge

(a) IA/1   1+   (b) IIA/2   2+
(c) IIIA/13   3+   (d) IVA/14   4+

73.   Ion   Ionic Charge

(a) Cs ion   1+   (b) Ga ion   3+
(c) O ion   2–   (d) I ion   1–

75.   Ion   Isoelectric

(a) Al^{3+} Ne   (b) Ca^{2+} Ar
(c) S^{2-} Ar   (d) N^{3-} Ne

77.   Ion   Electron Configuration

(a) Mg^{2+} $1s^2\,2s^2\,2p^6$ or [Ne]
(b) K^+ $1s^2\,2s^2\,2p^6\,3s^2\,3p^6$ or [Ar]
(c) Fe^{2+} $1s^2\,2s^2\,2p^6\,3s^2\,3p^6\,3d^6$ or [Ar] $3d^6$
(d) Cs^+ $1s^2\,2s^2\,2p^6\,3s^2\,3p^6\,4s^2\,3d^{10}\,4p^6\,5s^2\,4d^{10}\,5p^6$ or [Xe]

79.   Ion   Electron Configuration

(a) F^- $1s^2\,2s^2\,2p^6$ or [Ne]
(b) S^{2-} $1s^2\,2s^2\,2p^6\,3s^2\,3p^6$ or [Ar]
(c) N^{3-} $1s^2\,2s^2\,2p^6$ or [Ne]
(d) I^- $1s^2\,2s^2\,2p^6\,3s^2\,3p^6\,4s^2\,3d^{10}\,4p^6\,5s^2\,4d^{10}\,5p^6$ or [Xe]

General Exercises

81.   The element Mendeleev predicted and called ekaboron, we now call
scandium (Sc), which is named for Scandinavia its place of discovery.

83.   European   American   European   American

(a) Group IA   Group IA   (b) Group IB   Group IB
(c) Group IIIA Group IIIB   (d) Group IIIB Group IIIA

85. Predicted atomic radius of Fr: $0.266 + (0.266 - 0.248) = \sim 0.284$ nm

Predicted density of Fr: $1.87 + (1.87 - 1.53) = \sim 2.21$ g/mL

Predicted melting point of Fr: $28.4 - (38.9 - 28.4) = \sim 17.9°C$

87.

Element	Electron Configuration
(a) Sr	$[Kr] 5s^2$
(b) Ru	$[Kr] 5s^2 4d^6$
(c) Sb	$[Kr] 5s^2 4d^{10} 5p^3$
(d) Cs	$[Xe] 6s^1$

89. When an alkali metal atom loses one electron, it assumes a noble gas electron configuration. When an alkaline earth metal atom loses one electron, it does not assume a noble gas electron configuration. However, when an alkaline earth metal atom loses two electrons, it does attain a noble gas electron configuration and becomes very stable.

91. Although hydrogen and the Group IA/1 metals each have one valence electron, in hydrogen the electron is in the $1s$ sublevel and in the Group IA/1 metals the electron is in a higher s sublevel. As a result, in hydrogen the electron is closer to its nucleus than the valence electron in the Group IA/1 metals. Because the negatively charged electron is closer to the positively charged nucleus, it requires more energy to remove the electron in a hydrogen atom than in the other Group IA/1 metals.

Language of Chemistry

Section 7.1 *Classification of Compounds*

1.

	Compound	Classification		Compound	Classification
(a)	PbS	binary ionic	(b)	NH_3	binary molecular
(c)	$AgNO_3$	ternary ionic	(d)	$H_2SO_3(aq)$	ternary oxyacid
(e)	HBr(aq)	binary acid			

3.

	Ion	Classification		Ion	Classification
(a)	Br^-	monoatomic anion	(b)	NH_4^+	polyatomic cation
(c)	$C_2H_3O_2^-$	polyatomic anion	(d)	Hg^{2+}	monoatomic cation

Section 7.2 *Monoatomic Ions*

5.

	Monoatomic Cation	Systematic Name
(a)	K^+	potassium ion
(b)	Ba^{2+}	barium ion
(c)	Ag^+	silver ion
(d)	Cd^{2+}	cadmium ion

7.

	Monoatomic Cation	Stock System Name
(a)	Hg^{2+}	mercury(II) ion
(b)	Cu^{2+}	copper(II) ion
(c)	Fe^{2+}	iron(II) ion
(d)	Co^{3+}	cobalt(III) ion

9.

	Monoatomic Cation	Latin System Name
(a)	Cu^+	cuprous ion
(b)	Fe^{3+}	ferric ion
(c)	Sn^{2+}	stannous ion
(d)	Pb^{4+}	plumbic ion

11.   <u>Monoatomic Anion</u>   <u>Systematic Name</u>

    (a) F^- fluoride ion

    (b) I^- iodide ion

    (c) O^{2-} oxide ion

    (d) P^{3-} phosphide ion

Section 7.3 *Polyatomic Ions*

13.   <u>Polyatomic Ion</u>   <u>Systematic Name</u>

    (a) ClO^- hypochlorite ion

    (b) SO_3^{2-} sulfite ion

    (c) $C_2H_3O_2^-$ acetate ion

    (d) CO_3^{2-} carbonate ion

15.   <u>Polyatomic Ion</u>   <u>Chemical Formula</u>

    (a)   hydroxide ion OH^-

    (b)   nitrite ion NO_2^-

    (c)   dichromate ion $Cr_2O_7^{2-}$

    (d)   hydrogen carbonate ion HCO_3^-

Section 7.4 *Writing Chemical Formulas*

17.   <u>Ions</u>   <u>Binary Compound</u>

    (a) $Li^+ + Cl^-$ LiCl

    (b) $2\ Ag^+ + O^{2-}$ Ag_2O

    (c) $2\ Cr^{3+} + 3\ O^{2-}$ Cr_2O_3

    (d) $Sn^{4+} + 4\ I^-$ SnI_4

19.   <u>Ions</u>   <u>Ternary Compound</u>

    (a) $K^+ + NO_3^-$ KNO_3

    (b) $2\ NH_4^+ + Cr_2O_7^{2-}$ $(NH_4)_2Cr_2O_7$

    (c) $2\ Al^{3+} + 3\ SO_3^{2-}$ $Al_2(SO_3)_3$

    (d) $Cr^{3+} + 3\ ClO^-$ $Cr(ClO)_3$

21.   <u>Ions</u>   <u>Ternary Compound</u>

    (a) $Sr^{2+} + 2\ NO_2^-$ $Sr(NO_2)_2$

    (b) $Zn^{2+} + 2\ MnO_4^-$ $Zn(MnO_4)_2$

    (c) $Ca^{2+} + CrO_4^{2-}$ $CaCrO_4$

    (d) $Cr^{3+} + 3\ ClO_4^-$ $Cr(ClO_4)_3$

Section 7.5 *Binary Ionic Compounds*

23. | | Binary Ionic Compound | Systematic Name |
|---|---|---|
| (a) | MgO | magnesium oxide |
| (b) | $AgBr$ | silver bromide |
| (c) | $CdCl_2$ | cadmium chloride |
| (d) | Al_2S_3 | aluminum sulfide |

25. | | Binary Ionic Compound | Stock System Name |
|---|---|---|
| (a) | CuO | copper(II) oxide |
| (b) | FeO | iron(II) oxide |
| (c) | HgO | mercury(II) oxide |
| (d) | SnO | tin(II) oxide |

27. | | Binary Ionic Compound | Latin System Name |
|---|---|---|
| (a) | Cu_2O | cuprous oxide |
| (b) | Fe_2O_3 | ferric oxide |
| (c) | Hg_2O | mercurous oxide |
| (d) | SnO_2 | stannic oxide |

29. | | Binary Ionic Compound | Chemical Formula |
|---|---|---|
| (a) | rubidium chloride | $RbCl$ |
| (b) | sodium bromide | $NaBr$ |

31. | | Binary Ionic Compound | Chemical Formula |
|---|---|---|
| (a) | gallium nitride | GaN |
| (b) | aluminum arsenide | $AlAs$ |

Section 7.6 *Ternary Ionic Compounds*

33. | | Ternary Ionic Compound | Systematic Name |
|---|---|---|
| (a) | $LiMnO_4$ | lithium permanganate |
| (b) | $Sr(ClO_4)_2$ | strontium perchlorate |
| (c) | $CaCrO_4$ | calcium chromate |
| (d) | $Cd(CN)_2$ | cadmium cyanide |

35. | | Ternary Ionic Compound | Stock System Name |
|---|---|---|
| (a) | $CuSO_4$ | copper(II) sulfate |
| (b) | $FeCrO_4$ | iron(II) chromate |
| (c) | $Hg(NO_2)_2$ | mercury(II) nitrite |
| (d) | $Pb(C_2H_3O_2)_2$ | lead(II) acetate |

37. | Ternary Ionic Compound | Latin System Name |
|---|---|
| (a) Cu_2SO_4 | cuprous sulfate |
| (b) $Fe_2(CrO_4)_3$ | ferric chromate |
| (c) $Hg_2(NO_2)_2$ | mercurous nitrite |
| (d) $Pb(C_2H_3O_2)_4$ | plumbic acetate |

39. | Ternary Ionic Compound | Chemical Formula |
|---|---|
| (a) francium sulfate | Fr_2SO_4 |
| (b) sodium sulfite | Na_2SO_3 |

41. | Ternary Ionic Compound | Chemical Formula |
|---|---|
| (a) radium chlorate | $Ra(ClO_3)_2$ |
| (b) barium bromate | $Ba(BrO_3)_2$ |

Section 7.7 *Binary Molecular Compounds*

43. | Binary Molecular Compound | Systematic Name |
|---|---|
| (a) SO_3 | sulfur trioxide |
| (b) P_2O_3 | diphosphorus trioxide |
| (c) N_2O | dinitrogen oxide |
| (d) C_3O_2 | tricarbon dioxide |

45. | Binary Molecular Compound | Chemical Formula |
|---|---|
| (a) dinitrogen pentaoxide | N_2O_5 |
| (b) carbon tetrachloride | CCl_4 |
| (c) iodine monobromide | IBr |
| (d) dihydrogen sulfide | H_2S |

Section 7.8 *Binary Acids*

47. | Binary Acid | Systematic Name |
|---|---|
| (a) HBr(aq) | hydrobromic acid |
| (b) HI(aq) | hydroiodic acid |

Section 7.9 *Ternary Oxyacids*

49. | Ternary Oxyacid | Systematic Name |
|---|---|
| (a) $HClO_2(aq)$ | chlorous acid |
| (b) $H_3PO_4(aq)$ | phosphoric acid |

51. | Ternary Oxyacid | Chemical Formula |
|---|---|
| (a) acetic acid | $HC_2H_3O_2(aq)$ |
| (b) phosphorous acid | $H_3PO_3(aq)$ |

53.

	Ternary Oxyacid	Chemical Formula
(a)	chlorous acid	$HClO_2(aq)$
(b)	hypobromous acid	$HBrO(aq)$

General Exercises

55.

	Substance	Ionic Charge
(a)	iron metal atoms	0
(b)	ferrous ions	$2+$
(c)	iron(III) ions	$3+$
(d)	iron compounds	0

(*Note*: The total ionic charge on a compound is zero.)

57.

	Polyatomic Anion	Valence Electrons
(a)	$IO_4{}^{?-}$	$7 + 4(6) + (2\ e^- \text{ from charge}) = 33$
(b)	$SiO_3{}^{?-}$	$4 + 3(6) + (2\ e^- \text{ from charge}) = 24$

The polyatomic anion with a charge of 2– must be $SiO_3{}^{2-}$ because the total of valence electrons is an even number (24).

59.

Ions	F^-	O^{2-}	N^{3-}
Ag^+	AgF silver fluoride	Ag_2O silver oxide	Ag_3N silver nitride
$Hg_2{}^{2+}$	Hg_2F_2 mercury(I) fluoride	Hg_2O mercury(I) oxide	$(Hg_2)_3N_2$ mercury(I) nitride
Al^{3+}	AlF_3 aluminum fluoride	Al_2O_3 aluminum oxide	AlN aluminum nitride

61.

Ions	$MnO_4{}^-$	$SO_3{}^{2-}$	$PO_4{}^{3-}$
Cu^+	$CuMnO_4$ copper(I) permanganate	Cu_2SO_3 copper(I) sulfite	Cu_3PO_4 copper(I) phosphate
Cd^{2+}	$Cd(MnO_4)_2$ cadmium permanganate	$CdSO_3$ cadmium sulfite	$Cd_3(PO_4)_2$ cadmium phosphate
Cr^{3+}	$Cr(MnO_4)_3$ chromium(III) permanganate	$Cr_2(SO_3)_3$ chromium(III) sulfite	$CrPO_4$ chromium(III) phosphate

63.

	Compound	Suffix Ending			Compound	Suffix Ending
(a)	Na_2S	–ide		(b)	$H_2S(aq)$	–ic acid

65.

	Compound	Suffix Ending			Compound	Suffix Ending
(a)	Na_2SO_3	–ite		(b)	$H_2SO_3(aq)$	–ous acid

67.

	Compound	Suffix Ending			Compound	Suffix Ending
(a)	Na_2SO_4	–ate		(b)	$H_2SO_4(aq)$	–ic acid

69.

	Chemical Name	Chemical Formula
(a)	dihydrogen oxide (water)	H_2O
(b)	sodium hypochlorite (bleach)	$NaClO$
(c)	sodium hydroxide (caustic soda)	$NaOH$
(d)	sodium bicarbonate (baking soda)	$NaHCO_3$

71.

	Binary Compound	Systematic Name
(a)	BF_3	boron trifluoride
(b)	$SiCl_4$	silicon tetrachloride
(c)	As_2O_5	diarsenic pentaoxide
(d)	Sb_2O_3	diantimony trioxide

73.

Systematic Name	Chemical Formula
calcium acetate	$Ca(C_2H_3O_2)_2$

75. Given the chemical formula for lutetium chloride, $LuCl_3$, we can predict the formula for lawrencium chloride, $LrCl_3$, because the two metals are each at the end of the inner transition series.

Chemical Reactions

Section 8.1 *Evidence for Chemical Reactions*

1. (a) The production of a gas is evidence of a *chemical reaction*.
   (b) The formation of a precipitate is evidence of a *chemical reaction*.
   (c) A permanent color change is evidence of a *chemical reaction*.
   (d) A change in volume is not evidence of a chemical reaction, but rather a *physical change*.

3. (a) The release of light energy is evidence of a *chemical reaction*.
   (b) The release of heat energy is evidence of a *chemical reaction*.
   (c) A substance burning is evidence of a *chemical reaction*.
   (d) An explosion is evidence of a *chemical reaction*.

Section 8.2 *Writing Chemical Equations*

5. The following seven nonmetals occur naturally as diatomic molecules: H_2, N_2, O_2, F_2, Cl_2, Br_2, and I_2.

7. $2\,Fe(s) + 3\,Cl_2(g) \rightarrow 2\,FeCl_3(s)$

9. $ZnCO_3(s) \rightarrow ZnO(s) + CO_2(g)$

11. $Mg(s) + Co(NO_3)_2(aq) \rightarrow Mg(NO_3)_2(aq) + Co(s)$

13. $LiBr(aq) + AgNO_3(aq) \rightarrow AgBr(s) + LiNO_3(aq)$

15. $HC_2H_3O_2(aq) + KOH(aq) \rightarrow KC_2H_3O_2(aq) + H_2O(l)$

Section 8.3 *Balancing Chemical Equations*

17. The statements (a), (b), (c), and (d) are true; (e) is false.
    (e) If a coefficient is a fraction, multiply all the equation coefficients to clear the fraction and give whole numbers. Subscripts are never changed to balance a chemical equation. In fact, if an equation is difficult to balance, it is often because of an incorrect chemical formula.

19. (a) $4\,Co(s) + 3\,O_2(g) \rightarrow 2\,Co_2O_3(s)$
    (b) $2\,LiClO_3(s) \rightarrow 2\,LiCl(s) + 3\,O_2(g)$
    (c) $Cu(s) + 2\,AgC_2H_3O_2(aq) \rightarrow Cu(C_2H_3O_2)_2(aq) + 2\,Ag(s)$
    (d) $Pb(NO_3)_2(aq) + 2\,LiCl(aq) \rightarrow PbCl_2(s) + 2\,LiNO_3(aq)$
    (e) $3\,H_2SO_4(aq) + 2\,Al(OH)_3(aq) \rightarrow Al_2(SO_4)_3(aq) + 6\,H_2O(l)$

21. (a) $H_2CO_3(aq) + 2\,NH_4OH(aq) \rightarrow (NH_4)_2CO_3(aq) + 2\,HOH(l)$
    (b) $Hg_2(NO_3)_2(aq) + 2\,NaBr(aq) \rightarrow Hg_2Br_2(s) + 2\,NaNO_3(aq)$
    (c) $Mg(s) + 2\,HC_2H_3O_2(aq) \rightarrow Mg(C_2H_3O_2)_2(aq) + H_2(g)$
    (d) $2\,LiNO_3(s) \rightarrow 2\,LiNO_2(s) + O_2(g)$
    (e) $2\,Pb(s) + O_2(g) \rightarrow 2\,PbO(s)$

Section 8.4 *Classifying Chemical Reactions*

23. Refer to the chemical reactions in Exercise 19.
    (a) combination reaction
    (b) decomposition reaction
    (c) single–replacement reaction
    (d) double–replacement reaction
    (e) neutralization reaction

25. Refer to the chemical reactions in Exercise 21.
    (a) neutralization reaction
    (b) double–replacement reaction
    (c) single–replacement reaction
    (d) decomposition reaction
    (e) combination reaction

Section 8.5 *Combination Reactions*

General Form: $A + Z \rightarrow AZ$

27. metal + oxygen gas $\rightarrow$ metal oxide
    (a) $4\,Fe(s) + 3\,O_2(g) \rightarrow 2\,Fe_2O_3(s)$
    (b) $2\,Sn(s) + O_2(g) \rightarrow 2\,SnO(s)$

29. nonmetal + oxygen gas $\rightarrow$ nonmetal oxide
(a) $2\,C(s) + O_2(g) \rightarrow 2\,CO(g)$
(b) $4\,P(s) + 5\,O_2(g) \rightarrow 2\,P_2O_5(s)$

31. metal + nonmetal $\rightarrow$ ionic compound
(a) $2\,Cu(s) + Cl_2(g) \rightarrow 2\,CuCl(s)$
(b) $Co(s) + S(s) \rightarrow CoS(s)$

33. (a) $4\,Cr(s) + 3\,O_2(g) \rightarrow 2\,Cr_2O_3(s)$
(b) $2\,Cr(s) + N_2(g) \rightarrow 2\,CrN(s)$

35. (a) $4\,Li + O_2 \rightarrow 2\,Li_2O$
(b) $2\,Ca + O_2 \rightarrow 2\,CaO$

37. (a) $2\,Na + I_2 \rightarrow 2\,NaI$
(b) $3\,Ba + N_2 \rightarrow Ba_3N_2$

Section 8.6 *Decomposition Reactions*

General Form: AX $\rightarrow$ A + X

39. metal hydrogen carbonate $\rightarrow$ metal carbonate + water + carbon dioxide
(a) $2\,AgHCO_3(s) \rightarrow Ag_2CO_3(s) + H_2O(g) + CO_2(g)$
(b) $Ba(HCO_3)_2(s) \rightarrow BaCO_3(s) + H_2O(g) + CO_2(g)$

41. metal carbonate $\rightarrow$ metal oxide + carbon dioxide
(a) $K_2CO_3(s) \rightarrow K_2O(s) + CO_2(g)$
(b) $MnCO_3(s) \rightarrow MnO(s) + CO_2(g)$

43. oxygen–containing compounds $\rightarrow$ oxygen gas
(a) $Ca(NO_3)_2(s) \rightarrow Ca(NO_2)_2(s) + O_2(g)$
(b) $2\,Ag_2SO_4(s) \rightarrow 2\,Ag_2SO_3(s) + O_2(g)$

45. metal hydrogen carbonate $\rightarrow$ metal carbonate + water + carbon dioxide
(a) $2\,KHCO_3(s) \rightarrow K_2CO_3(s) + H_2O(g) + CO_2(g)$
(b) $Zn(HCO_3)_2(s) \rightarrow ZnCO_3(s) + H_2O(g) + CO_2(g)$

47. oxygen–containing compounds $\rightarrow$ oxygen gas
(a) $2\,NaClO_3(s) \rightarrow 2\,NaCl(s) + 3\,O_2(g)$
(b) $Ca(NO_3)_2(s) \rightarrow Ca(NO_2)_2(s) + O_2(g)$

Section 8.7 *The Activity Series Concept*

49.

	Element	Solution	Observation
(a)	Hg	$Fe(NO_3)_2(aq)$	no reaction; Fe > Hg
(b)	Zn	$Fe(NO_3)_2(aq)$	reaction; Zn > Fe
(c)	Cd	$Fe(NO_3)_2(aq)$	no reaction; Fe > Cd
(d)	Mg	$Fe(NO_3)_2(aq)$	reaction; Mg > Fe

51.

	Element	Solution	Observation
(a)	Ni	$HCl(aq)$	reaction; Ni > (H)
(b)	Zn	$HCl(aq)$	reaction; Zn > (H)
(c)	Cu	$HCl(aq)$	no reaction; Cu < (H)
(d)	Al	$HCl(aq)$	reaction; Al > (H)

53.

	Element	Solution	Observation
(a)	Li	$H_2O(l)$	reaction; Li is an active metal
(b)	Mg	$H_2O(l)$	no reaction; Mg is not an active metal
(c)	Ca	$H_2O(l)$	reaction; Ca is an active metal
(d)	Al	$H_2O(l)$	no reaction; Al is not an active metal

Section 8.8 *Single–Replacement Reactions*

General Form: A + BZ → AZ + B
or: $A + BZ → NR$

55. $metal_1$ + aqueous $solution_1$ → $metal_2$ + aqueous $solution_2$
(a) $Cu(s) + Al(NO_3)_3(aq) → NR$
(b) $2\,Al(s) + 3\,Cu(NO_3)_2(aq) → 3\,Cu(s) + 2\,Al(NO_3)_3(aq)$

57. $metal_1$ + aqueous $solution_1$ → $metal_2$ + aqueous $solution_2$
(a) $Ni(s) + Pb(C_2H_3O_2)_2(aq) → Pb(s) + Ni(C_2H_3O_2)_2(aq)$
(b) $Pb(s) + Ni(C_2H_3O_2)_2(aq) → NR$

59. metal + aqueous acid → aqueous solution + hydrogen gas
(a) $Mg(s) + 2\,HCl(aq) → MgCl_2(s) + H_2(g)$
(b) $Mn(s) + 2\,HNO_3(aq) → Mn(NO_3)_2(aq) + H_2(g)$

61. metal + water → metal hydroxide + hydrogen gas
(a) $2\,Li(s) + 2\,H_2O(l) → 2\,LiOH(aq) + H_2(g)$
(b) $Ba(s) + 2\,H_2O(l) → Ba(OH)_2(aq) + H_2(g)$

63. $metal_1$ + aqueous $solution_1$ → $metal_2$ + aqueous $solution_2$
(a) $Zn(s) + Pb(NO_3)_2(aq) → Zn(NO_3)_2(aq) + Pb(s)$
(b) $Cd(s) + Fe(NO_3)_2(aq) → NR$

65. metal + aqueous acid → aqueous solution + hydrogen gas
 (a) $Zn(s) + 2 HNO_3(aq) \rightarrow Zn(NO_3)_2(aq) + H_2(g)$
 (b) $Cd(s) + 2 HNO_3(aq) \rightarrow Cd(NO_3)_2(aq) + H_2(g)$

67. metal + water → metal hydroxide + hydrogen gas
 (a) $2 K(s) + 2 H_2O(l) \rightarrow 2 KOH(aq) + H_2(g)$
 (b) $Ba(s) + 2 H_2O(l) \rightarrow Ba(OH)_2(aq) + H_2(g)$

Section 8.9 *Solubility Rules*

69. (a) $Co(OH)_2$ insoluble (b) $FeSO_4$ soluble
 (c) $SnCrO_4$ insoluble (d) $Pb(C_2H_3O_2)_2$ soluble

71. (a) Hg_2Cl_2 insoluble (b) $HgCl_2$ soluble
 (c) $AgBr$ insoluble (d) PbI_2 insoluble

Section 8.10 *Double–Replacement Reactions*

General Form: AX + BZ → AZ + BX
 or: $AX + BZ \rightarrow NR$

73. aqueous solution$_1$ + aqueous solution$_2$ → precipitate + aqueous solution$_3$
 (a) $ZnCl_2(aq) + 2 NH_4OH(aq) \rightarrow Zn(OH)_2(s) + 2 NH_4Cl(aq)$
 (b) $NiSO_4(aq) + Hg_2(NO_3)_2(aq) \rightarrow Hg_2SO_4(s) + Ni(NO_3)_2(aq)$

75. aqueous solution$_1$ + aqueous solution$_2$ → precipitate + aqueous solution$_3$
 (a) $MgSO_4(aq) + BaCl_2(aq) \rightarrow BaSO_4(s) + MgCl_2(aq)$
 (b) $2 AlBr_3(aq) + 3 Na_2CO_3(aq) \rightarrow Al_2(CO_3)_3(s) + 6 NaBr(aq)$

Section 8.11 *Neutralization Reactions*

General Form: HX + BOH → BX + HOH

77. aqueous acid + aqueous base → aqueous salt + water
 (a) $NaOH(aq) + HNO_3(aq) \rightarrow NaNO_3(aq) + HOH(l)$

 aqueous acid + aqueous base → precipitate + water
 (b) $3 Ba(OH)_2(aq) + 2 H_3PO_4(aq) \rightarrow Ba_3(PO_4)_2(s) + 6 HOH(l)$

79. aqueous acid + aqueous base → aqueous salt + water
 (a) $2 HF(aq) + Ca(OH)_2(aq) \rightarrow CaF_2(aq) + 2 HOH(l)$
 (b) $H_2SO_4(aq) + 2 LiOH(aq) \rightarrow Li_2SO_4(aq) + 2 HOH(l)$

General Exercises

81.   (a) $3\,Fe(s) + 4\,H_2O(g) \rightarrow Fe_3O_4(s) + 4\,H_2(g)$
       (b) $4\,FeS(s) + 7\,O_2(g) \rightarrow 2\,Fe_2O_3(s) + 4\,SO_2(g)$

83.   (a) $F_2(g) + 2\,NaBr(aq) \rightarrow Br_2(l) + 2\,NaF(aq)$
       (b) $Sb_2S_3(s) + 6\,HCl(aq) \rightarrow 2\,SbCl_3(aq) + 3\,H_2S(aq)$

85.   (a) $CH_4(g) + 2\,O_2(g) \rightarrow CO_2(g) + 2\,H_2O(g)$
       (b) $C_3H_8(g) + 5\,O_2(g) \rightarrow 3\,CO_2(g) + 4\,H_2O(g)$

87.   Industrial Preparation of Chlorine:
      $4\,HCl(g) + O_2(g) \rightarrow 2\,Cl_2(g) + 2\,H_2O(g)$

89.   Contact Process for Sulfuric Acid:
      (1) $S(s) + O_2(g) \rightarrow SO_2(g)$

      (2) $2\,SO_2(g) + O_2(g) \xrightarrow{Pt} 2\,SO_3(g)$

      (3) $SO_3(g) + H_2O(l) \rightarrow H_2SO_4(aq)$

The Mole Concept

Section 9.1 *Avogadro's Number*

1.

	Element	Average Mass		Element	Average Mass
(a)	H	1.01 amu	(b)	Li	6.94 amu
(c)	C	12.01 amu	(d)	P	30.97 amu

3.

	Element	Average Mass		Element	Average Mass
(a)	H	1.01 g	(b)	Li	6.94 g
(c)	C	12.01 g	(d)	P	30.97 g

Section 9.2 *Mole Calculations I*

5. (a) 1 mole Mn atoms $= 6.02 \times 10^{23}$ atoms Mn
   (b) 1 mole $Mn(NO_3)_2$ formula units $= 6.02 \times 10^{23}$ formula units $Mn(NO_3)_2$

7. (a) 6.02×10^{23} atoms Cu $= 1$ mol Cu atoms
   (b) 6.02×10^{23} formula units $CuSO_4 = 1$ mol $CuSO_4$ formula units

9. (a) $0.335 \; \cancel{\text{mol Ti}} \times \dfrac{6.02 \times 10^{23} \text{ atoms Ti}}{1 \; \cancel{\text{mol Ti}}} = 2.02 \times 10^{23}$ atoms Ti

   (b) $0.112 \; \cancel{\text{mol CO}_2} \times \dfrac{6.02 \times 10^{23} \text{ molecules CO}_2}{1 \; \cancel{\text{mol CO}_2}} = 6.74 \times 10^{22}$ molecules CO_2

   (c) $1.935 \; \cancel{\text{mol ZnCl}_2} \times \dfrac{6.02 \times 10^{23} \text{ formula units ZnCl}_2}{1 \; \cancel{\text{mol ZnCl}_2}}$

   $= 1.16 \times 10^{24}$ formula units $ZnCl_2$

11. (a) $4.15 \times 10^{22} \text{ atoms Fe} \times \dfrac{1 \text{ mol Fe}}{6.02 \times 10^{23} \text{ atoms Fe}} = 0.0689 \text{ mol Fe}$

(b) $3.31 \times 10^{21} \text{ molecules Br}_2 \times \dfrac{1 \text{ mol Br}_2}{6.02 \times 10^{23} \text{ molecules Br}_2} = 0.00550 \text{ mol Br}_2$

(c) $4.19 \times 10^{20} \text{ formula units Cd(NO}_3)_2 \times \dfrac{1 \text{ mol Cd(NO}_3)_2}{6.02 \times 10^{23} \text{ formula units Cd(NO}_3)_2}$

$$= 6.96 \times 10^{-4} \text{ mol Cd(NO}_3)_2$$

Section 9.3 *Molar Mass*

13.

	Element	Molar Mass
(a)	Hg	200.59 g/mol
(b)	Si	28.09 g/mol
(c)	Br$_2$	2(79.90) = 159.80 g/mol
(d)	P$_4$	4(30.97) = 123.88 g/mol

15.

	Compound	Molar Mass	
(a)	BaF$_2$	137.33 g Ba + 2(19.00 g F)	= 175.33 g/mol
(b)	K$_2$S	2(39.10 g K) + 32.07 g S	= 110.27 g/mol
(c)	Fe(C$_2$H$_3$O$_2$)$_3$	55.85 g Fe + 6(12.01 g C) + 9(1.01 g H) + 6(16.00 g O)	= 233.00 g/mol
(d)	Sr$_3$(PO$_4$)$_2$	3(87.62 g Sr) + 2(30.97 g P) + 8(16.00 g O)	= 452.80 g/mol

Section 9.4 *Mole Calculations II*

17. (a) MM of Hg = 200.59 g/mol

$$2.95 \times 10^{23} \text{ atoms Hg} \times \dfrac{1 \text{ mol Hg}}{6.02 \times 10^{23} \text{ atoms Hg}} \times \dfrac{200.59 \text{ g Hg}}{1 \text{ mol Hg}}$$

$$= 98.3 \text{ g Hg}$$

(b) MM of N$_2$ = 2(14.01 g N) = 28.02 g/mol

$$1.16 \times 10^{22} \text{ molecules N}_2 \times \dfrac{1 \text{ mol N}_2}{6.02 \times 10^{23} \text{ molecules N}_2} \times \dfrac{28.02 \text{ g N}_2}{1 \text{ mol N}_2}$$

$$= 0.540 \text{ g N}_2$$

(c) MM of $BaCl_2$ = 137.33 g Ba + 2(35.45 g Cl) = 208.23 g/mol

$$5.05 \times 10^{21} \text{ formula units BaCl}_2 \times \frac{1 \text{ mol BaCl}_2}{6.02 \times 10^{23} \text{ formula units BaCl}_2} \times$$

$$\frac{208.23 \text{ g BaCl}_2}{1 \text{ mol BaCl}_2} = 1.75 \text{ g BaCl}_2$$

19. (a) MM of K = 39.10 g/mol

$$1.50 \text{ g K} \times \frac{1 \text{ mol K}}{39.10 \text{ g K}} \times \frac{6.02 \times 10^{23} \text{ atoms K}}{1 \text{ mol K}}$$

$$= 2.31 \times 10^{22} \text{ atoms K}$$

(b) MM of O_2 = 2(16.00 g O) = 32.00 g/mol

$$0.470 \text{ g O}_2 \times \frac{1 \text{ mol O}_2}{32.00 \text{ g O}_2} \times \frac{6.02 \times 10^{23} \text{ molecules O}_2}{1 \text{ mol O}_2}$$

$$= 8.84 \times 10^{21} \text{ molecules O}_2$$

(c) MM of $AgClO_3$ = 107.87 g Ag + 35.45 g Cl + 3(16.00 g O) = 191.32 g/mol

$$0.555 \text{ g AgClO}_3 \times \frac{1 \text{ mol AgClO}_3}{191.32 \text{ g AgClO}_3} \times \frac{6.02 \times 10^{23} \text{ formula units AgClO}_3}{1 \text{ mol AgClO}_3}$$

$$= 1.75 \times 10^{21} \text{ formula units AgClO}_3$$

21. (a) $\dfrac{9.01 \text{ g Be}}{1 \text{ mol Be}} \times \dfrac{1 \text{ mol Be}}{6.02 \times 10^{23} \text{ atoms Be}} = 1.50 \times 10^{-23} \text{ g/atom}$

(b) $\dfrac{22.99 \text{ g Na}}{1 \text{ mol Na}} \times \dfrac{1 \text{ mol Na}}{6.02 \times 10^{23} \text{ atoms Na}} = 3.82 \times 10^{-23} \text{ g/atom}$

(c) $\dfrac{58.93 \text{ g Co}}{1 \text{ mol Co}} \times \dfrac{1 \text{ mol Co}}{6.02 \times 10^{23} \text{ atoms Co}} = 9.79 \times 10^{-23} \text{ g/atom}$

(d) $\dfrac{74.92 \text{ g As}}{1 \text{ mol As}} \times \dfrac{1 \text{ mol As}}{6.02 \times 10^{23} \text{ atoms As}} = 1.24 \times 10^{-22} \text{ g/atom}$

Section 9.5 *Molar Volume*

23. Standard conditions: 0°C and 1 atmosphere (1 atm)

25. (a) MM of Ne = 20.18 g/mol

Density of Ne (at STP): $\dfrac{20.18 \text{ g}}{1 \text{ mol}} \times \dfrac{1 \text{ mol}}{22.4 \text{ L}} = 0.901 \text{ g/L}$

(b) MM of Cl_2 = 2(35.45 g Cl) = 70.90 g/mol

Density of Cl_2 (at STP): $\dfrac{70.90 \text{ g}}{1 \text{ mol}} \times \dfrac{1 \text{ mol}}{22.4 \text{ L}} = 3.17 \text{ g/L}$

(c) MM of NO_2 = 14.01 g N + 2(16.00 g O) = 46.01 g/mol

Density of NO_2 (at STP): $\dfrac{46.01 \text{ g}}{1 \text{ mol}} \times \dfrac{1 \text{ mol}}{22.4 \text{ L}} = 2.05 \text{ g/L}$

(d) MM of HI = 1.01 g H + 126.90 g I = 127.91 g/mol

Density of HI (at STP): $\dfrac{127.91 \text{ g}}{1 \text{ mol}} \times \dfrac{1 \text{ mol}}{22.4 \text{ L}} = 5.71 \text{ g/L}$

27. (a) MM of ethane: $\dfrac{22.4 \text{ L}}{1 \text{ mol}} \times \dfrac{1.34 \text{ g}}{1 \text{ L}} = 30.0 \text{ g/mol}$

(b) MM of diborane: $\dfrac{22.4 \text{ L}}{\text{mol}} \times \dfrac{1.23 \text{ g}}{1 \text{ L}} = 27.6 \text{ g/mol}$

(c) MM of Freon–12: $\dfrac{22.4 \text{ L}}{\text{mol}} \times \dfrac{5.40 \text{ g}}{1 \text{ L}} = 121 \text{ g/mol}$

(d) MM of nitrous oxide: $\dfrac{22.4 \text{ L}}{\text{mol}} \times \dfrac{2.05 \text{ g}}{1 \text{ L}} = 45.9 \text{ g/mol}$

29.

Gas	Molecules	Mass	Volume @ STP
fluorine, F_2	6.02×10^{23}	38.00 g	22.4 L
hydrogen fluoride, HF	6.02×10^{23}	20.01 g	22.4 L
silicon tetrafluoride, SiF_4	6.02×10^{23}	104.09 g	22.4 L
oxygen difluoride, OF_2	6.02×10^{23}	54.00 g	22.4 L

Section 9.6 *Mole Calculations III*

31. (a) MM of He = 4.00 g/mol

$$0.250 \text{ g He} \times \frac{1 \text{ mol He}}{4.00 \text{ g He}} \times \frac{22.4 \text{ L He}}{1 \text{ mol He}} = 1.40 \text{ L He}$$

(b) MM of N_2 = 2(14.01 g N) = 28.02 g/mol

$$5.05 \text{ g N}_2 \times \frac{1 \text{ mol N}_2}{28.02 \text{ g N}_2} \times \frac{22.4 \text{ L N}_2}{1 \text{ mol N}_2} = 4.04 \text{ L N}_2$$

33. (a) MM of H_2S = 2(1.01 g H) + 32.07 g S = 34.09 g/mol

$$1.05 \text{ L H}_2\text{S} \times \frac{1 \text{ mol H}_2\text{S}}{22.4 \text{ L H}_2\text{S}} \times \frac{34.09 \text{ g H}_2\text{S}}{1 \text{ mol H}_2\text{S}} = 1.60 \text{ g H}_2\text{S}$$

(b) MM of N_2O_3 = 2(14.01 g N) + 3(16.00 g O) = 76.02 g/mol

$$5.33 \text{ L N}_2\text{O}_3 \times \frac{1 \text{ mol N}_2\text{O}_3}{22.4 \text{ L N}_2\text{O}_3} \times \frac{76.02 \text{ g N}_2\text{O}_3}{1 \text{ mol N}_2\text{O}_3} = 18.1 \text{ g N}_2\text{O}_3$$

35. (a) $100.0 \text{ mL H}_2 \times \frac{1 \text{ L H}_2}{1000 \text{ mL H}_2} \times \frac{1 \text{ mol H}_2}{22.4 \text{ L H}_2} \times \frac{6.02 \times 10^{23} \text{ molecules H}_2}{1 \text{ mol H}_2}$

$$= 2.69 \times 10^{21} \text{ molecules H}_2$$

(b) $70.5 \text{ mL NH}_3 \times \frac{1 \text{ L NH}_3}{1000 \text{ mL NH}_3} \times \frac{1 \text{ mol NH}_3}{22.4 \text{ L NH}_3} \times \frac{6.02 \times 10^{23} \text{ molecules NH}_3}{1 \text{ mol NH}_3}$

$$= 1.89 \times 10^{21} \text{ molecules NH}_3$$

37.

Gas	Molecules	Atoms	Mass	Volume @ STP
N_2	1.35×10^{23}	2.70×10^{23}	6.28 g	5.02 L
NO_2	1.35×10^{23}	4.06×10^{23}	10.3 g	5.02 L
NO	1.35×10^{23}	2.71×10^{23}	6.75 g	5.02 L
N_2O_4	1.35×10^{23}	8.10×10^{23}	20.6 g	5.02 L

Section 9.7 *Percent Composition*

39. $\text{MM of } C_7H_6O_3 \quad = 7(12.01 \text{ g C}) + 6(1.01 \text{ g H}) + 3(16.00 \text{ g O})$

$\qquad\qquad\qquad = 84.07 \text{ g C} + 6.06 \text{ g H} + 48.00 \text{ g O}$

$\qquad\qquad\qquad = 138.13 \text{ g/mol}$

$$\frac{84.07 \text{ g C}}{138.13 \text{ g } C_7H_6O_3} \times 100 = 60.86\% \text{ C}$$

$$\frac{6.06 \text{ g H}}{138.13 \text{ g } C_7H_6O_3} \times 100 = 4.39\% \text{ H}$$

$$\frac{48.00 \text{ g O}}{138.13 \text{ g } C_7H_6O_3} \times 100 = 34.75\% \text{ O}$$

41. $\text{MM of } C_{17}H_{21}NO_4 = 17(12.01 \text{ g C}) + 21(1.01 \text{ g H}) + 14.01 \text{ g N} + 4(16.00 \text{ g O})$

$\qquad\qquad\qquad = 204.17 \text{ g C} + 21.21 \text{ g H} + 14.01 \text{ g N} + 64.00 \text{ g O}$

$\qquad\qquad\qquad = 303.39 \text{ g/mol}$

$$\frac{204.17 \text{ g C}}{303.39 \text{ g } C_{17}H_{21}NO_4} \times 100 = 67.296\% \text{ C}$$

$$\frac{21.21 \text{ g H}}{303.39 \text{ g } C_{17}H_{21}NO_4} \times 100 = 6.991\% \text{ H}$$

$$\frac{14.01 \text{ g N}}{303.39 \text{ g } C_{17}H_{21}NO_4} \times 100 = 4.618\% \text{ N}$$

$$\frac{64.00 \text{ g O}}{303.39 \text{ g } C_{17}H_{21}NO_4} \times 100 = 21.09\% \text{ O}$$

43. $\text{MM of } C_4H_8SCl_2 \quad = 4(12.01 \text{ g C}) + 8(1.01 \text{ g H}) + 32.07 \text{ g S} + 2(35.45 \text{ g Cl})$

$\qquad\qquad\qquad = 48.04 \text{ g C} + 8.08 \text{ g H} + 32.07 \text{ g S} + 70.90 \text{ g Cl}$

$\qquad\qquad\qquad = 159.09 \text{ g/mol}$

$$\frac{48.04 \text{ g C}}{159.09 \text{ g } C_4H_8SCl_2} \times 100 = 30.20\% \text{ C}$$

$$\frac{8.08 \text{ g H}}{159.09 \text{ g } C_4H_8SCl_2} \times 100 = 5.08\% \text{ H}$$

$$\frac{32.07 \text{ g S}}{159.09 \text{ g } C_4H_8SCl_2} \times 100 = 20.16\% \text{ S}$$

$$\frac{70.90 \text{ g Cl}}{159.09 \text{ g } C_4H_8SCl_2} \times 100 = 44.57\% \text{ Cl}$$

45. MM of $NaC_5H_8NO_4$

$$= 22.99\,g\,Na + 5(12.01\,g\,C) + 8(1.01\,g\,H) + 14.01\,g\,N + 4(16.00\,g\,O)$$

$$= 22.99\,g\,Na + 60.05\,g\,C + 8.08\,g\,H + 14.01\,g\,N + 64.00\,g\,O$$

$$= 169.13\ g/mol$$

$$\frac{22.99\ g\ Na}{169.13\ g\ NaC_5H_8NO_4} \times 100 = 13.59\%\ Na$$

$$\frac{60.05\ g\ C}{169.13\ g\ NaC_5H_8NO_4} \times 100 = 35.51\%\ C$$

$$\frac{8.08\ g\ H}{169.13\ g\ NaC_5H_8NO_4} \times 100 = 4.78\%\ H$$

$$\frac{14.01\ g\ N}{169.13\ g\ NaC_5H_8NO_4} \times 100 = 8.284\%\ N$$

$$\frac{64.00\,g\ O}{169.13\ g\ NaC_5H_8NO_4} \times 100 = 37.84\%\ O$$

Section 9.8 *Empirical Formula*

47. $0.500\ \cancel{g\ Sn} \times \dfrac{1\ mol\ Sn}{118.71\ \cancel{g\ Sn}} = 0.00421\ mol\ Sn$

$0.635\ g\ Sn_xO_y - 0.500\ g\ Sn = 0.135\ g\ O$

$0.135\ \cancel{g\ O} \times \dfrac{1\ mol\ O}{16.00\ \cancel{g\ O}} = 0.00844\ mol\ O$

$\dfrac{Sn_{0.00421}}{0.00421}\ \dfrac{O_{0.00844}}{0.00421} = Sn_{1.00}O_{2.00}$ The empirical formula is SnO_2.

49. $1.435\ \cancel{g\ Hg} \times \dfrac{1\ mol\ Hg}{200.59\ \cancel{g\ Hg}} = 0.007154\ mol\ Hg$

$1.550\ g\ Hg_xO_y - 1.435\ g\ Hg = 0.115\ g\ O$

$0.115\ \cancel{g\ O} \times \dfrac{1\ mol\ O}{16.00\ \cancel{g\ O}} = 0.00719\ mol\ O$

$\dfrac{Hg_{0.007154}}{0.007154}\ \dfrac{O_{0.00719}}{0.007154} = Hg_{1.000}O_{1.01}$ The empirical formula is HgO.

51. $1.115 \text{ g Co} \times \dfrac{1 \text{ mol Co}}{58.93 \text{ g Co}} = 0.0189 \text{ mol Co}$

$2.025 \text{ g Co}_x\text{S}_y - 1.115 \text{ g Co} = 0.910 \text{ g S}$

$0.910 \text{ g S} \times \dfrac{1 \text{ mol S}}{32.07 \text{ g S}} = 0.0284 \text{ mol S}$

$\dfrac{\text{Co } 0.0189}{0.0189} \quad \dfrac{\text{S } 0.0284}{0.0189} = \text{Co}_{1.00}\text{S}_{1.50}$ The empirical formula is Co_2S_3.

53. (a) $59.1 \text{ g Mn} \times \dfrac{1 \text{ mol Mn}}{54.94 \text{ g Mn}} = 1.08 \text{ mol Mn}$

$40.9 \text{ g F} \times \dfrac{1 \text{ mol F}}{19.00 \text{ g F}} = 2.15 \text{ mol F}$

$\dfrac{\text{Mn } 1.08}{1.08} \quad \dfrac{\text{F } 2.15}{1.08} = \text{Mn}_{1.00}\text{F}_{1.99}$ The empirical formula is MnF_2.

(b) $64.1 \text{ g Cu} \times \dfrac{1 \text{ mol Cu}}{63.55 \text{ g Cu}} = 1.01 \text{ mol Cu}$

$35.9 \text{ g Cl} \times \dfrac{1 \text{ mol Cl}}{35.45 \text{ g Cl}} = 1.01 \text{ mol Cl}$

$\dfrac{\text{Cu } 1.01}{1.01} \quad \dfrac{\text{Cl}_{1.01}}{1.01} = \text{Cu}_{1.00}\text{Cl}_{1.00}$ The empirical formula is $CuCl$.

(c) $42.6 \text{ g Sn} \times \dfrac{1 \text{ mol Sn}}{118.71 \text{ g Sn}} = 0.359 \text{ mol Sn}$

$57.4 \text{ g Br} \times \dfrac{1 \text{ mol Br}}{79.90 \text{ g Br}} = 0.718 \text{ mol Br}$

$\dfrac{\text{Sn } 0.359}{0.359} \quad \dfrac{\text{Br } 0.718}{0.359} = \text{Sn}_{1.00}\text{Br}_{2.00}$ The empirical formula is $SnBr_2$.

55. $18.25 \text{ g C} \times \dfrac{1 \text{ mol C}}{12.01 \text{ g C}} = 1.52 \text{ mol C}$

$0.77 \text{ g H} \times \dfrac{1 \text{ mol H}}{1.01 \text{ g H}} = 0.76 \text{ mol H}$

$80.99 \text{ g Cl} \times \dfrac{1 \text{ mol Cl}}{35.45 \text{ g Cl}} = 2.28 \text{ mol Cl}$

$\dfrac{\text{C } 1.52}{0.76} \quad \dfrac{\text{H } 0.76}{0.76} \quad \dfrac{\text{Cl } 2.28}{0.76} = \text{C}_{2.0}\text{H}_{1.0}\text{Cl}_{3.0}$ The empirical formula is C_2HCl_3.

Section 9.9 *Molecular Formula*

57. MM of $C_9H_8O_4$ $= 9(12.01 \text{ g C}) + 8(1.01 \text{ g H}) + 4(16.00 \text{ g O})$
$= 108.19 \text{ g C} + 8.08 \text{ g H} + 64.00 \text{ g O}$
$= 180.17 \text{ g/mol}$

Aspirin: $\dfrac{(C_9H_8O_4)_n}{C_9H_8O_4} = \dfrac{180 \text{ g/mol}}{180.17 \text{ g/mol}}$ $n \approx 1$

The molecular formula of aspirin is $(C_9H_8O_4)_1$ or $C_9H_8O_4$.

59. MM of $C_3H_5O_2$ $= 3(12.01 \text{ g C}) + 5(1.01 \text{ g H}) + 2(16.00 \text{ g O})$
$= 36.03 \text{ g C} + 5.05 \text{ g H} + 32.00 \text{ g O}$
$= 73.08 \text{ g/mol}$

Adipic acid: $\dfrac{(C_3H_5O_2)_n}{C_3H_5O_2} = \dfrac{147 \text{ g/mol}}{73.08 \text{ g/mol}}$ $n \approx 2$

The molecular formula of adipic acid is $(C_3H_5O_2)_2$ or $C_6H_{10}O_4$.

61. <u>Empirical Formula</u>

$38.7 \text{ g C} \times \dfrac{1 \text{ mol C}}{12.01 \text{ g C}} = 3.22 \text{ mol C}$

$9.74 \text{ g H} \times \dfrac{1 \text{ mol H}}{1.01 \text{ g H}} = 9.64 \text{ mol H}$

$51.6 \text{ g O} \times \dfrac{1 \text{ mol O}}{16.00 \text{ g O}} = 3.23 \text{ mol O}$

$C_{\frac{3.22}{3.22}} H_{\frac{9.64}{3.22}} O_{\frac{3.23}{3.22}} = C_{1.00}H_{2.99}O_{1.00}$ The empirical formula is CH_3O.

<u>Molecular Formula</u>

MM of CH_3O $= 12.01 \text{ g C} + 3(1.01 \text{ g H}) + 16.00 \text{ g O}$
$= 12.01 \text{ g C} + 3.03 \text{ g H} + 16.00 \text{ g O}$
$= 31.04 \text{ g/mol}$

Ethylene glycol: $\dfrac{(CH_3O)_n}{CH_3O} = \dfrac{62 \text{ g/mol}}{31.04 \text{ g/mol}}$ $n \approx 2$

The molecular formula of ethylene glycol is $(CH_3O)_2$ or $C_2H_6O_2$.

63. Empirical Formula

$$24.8 \text{ g C} \times \frac{1 \text{ mol C}}{12.01 \text{ g C}} = 2.06 \text{ mol C}$$

$$2.08 \text{ g H} \times \frac{1 \text{ mol H}}{1.01 \text{ g H}} = 2.06 \text{ mol H}$$

$$73.1 \text{ g Cl} \times \frac{1 \text{ mol Cl}}{35.45 \text{ g Cl}} = 2.06 \text{ mol Cl}$$

$$\text{C} \frac{2.06}{2.06} \text{ H} \frac{2.06}{2.06} \text{ Cl} \frac{2.06}{2.06} = C_{1.00}H_{1.00}Cl_{1.00}$$ The empirical formula is $CHCl$.

Molecular Formula

MM of CHCl $\quad = 12.01 \text{ g C} + 1.01 \text{ g H} + 35.45 \text{ g Cl}$
$\qquad = 48.47 \text{ g/mol}$

Lindane: $\quad \dfrac{(CHCl)_n}{CHCl} = \dfrac{290 \text{ g/mol}}{48.47 \text{ g/mol}} \qquad n \approx 6$

The molecular formula of lindane is $(CHCl)_6$ or $C_6H_6Cl_6$.

65. Empirical Formula

$$74.0 \text{ g C} \times \frac{1 \text{ mol C}}{12.01 \text{ g C}} = 6.16 \text{ mol C}$$

$$8.70 \text{ g H} \times \frac{1 \text{ mol H}}{1.01 \text{ g H}} = 8.61 \text{ mol H}$$

$$17.3 \text{ g N} \times \frac{1 \text{ mol N}}{14.01 \text{ g N}} = 1.23 \text{ mol N}$$

$$\text{C} \frac{6.16}{1.23} \text{ H} \frac{8.61}{1.23} \text{ N} \frac{1.23}{1.23} = C_{5.01}H_{7.00}N_{1.00}$$ The empirical formula is C_5H_7N.

Molecular Formula

MM of $C_5H_7N \quad = 5(12.01 \text{ g C}) + 7(1.01 \text{ g H}) + 14.01 \text{ g N}$
$\qquad = 60.05 \text{ g C} + 7.07 \text{ g H} + 14.01 \text{ g N}$
$\qquad = 81.13 \text{ g/mol}$

Nicotine: $\quad \dfrac{(C_5H_7N)_n}{C_5H_7N} = \dfrac{160 \text{ g/mol}}{81.13 \text{ g/mol}} \qquad n \approx 2$

The molecular formula of nicotine is $(C_5H_7N)_2$ or $C_{10}H_{14}N_2$.

General Exercises

67. $1 \text{ Faraday} = 1 \text{ mole of electrons} = 6.02 \times 10^{23} \text{ electrons}$

69. $5 \text{ g Ni} \times \dfrac{1 \text{ mol Ni}}{58.69 \text{ g Ni}} \times \dfrac{6.02 \times 10^{23} \text{ atoms Ni}}{1 \text{ mol Ni}} = 5 \times 10^{22} \text{ atoms Ni}$

$5 \times 10^{22} \text{ atoms Ni} >>> 1 \times 10^{15} \text{ red blood cells}$

Thus, the number of Ni atoms in a 5–g nickel coin is 50,000,000 times greater than the number of red blood cells in 50,000 people!

71. $1 \text{ mol furry moles} \times \dfrac{6.02 \times 10^{23} \text{ furry moles}}{1 \text{ mol furry moles}} \times \dfrac{100 \text{ g}}{1 \text{ furry mole}} \times \dfrac{1 \text{ kg}}{1000 \text{ g}}$

$$= 6 \times 10^{22} \text{ kg}$$

$6 \times 10^{24} \text{ kg (Earth)} >> 6 \times 10^{22} \text{ kg (1 mol furry moles)}$
Thus, the Earth weighs about 100 times more than a mole of moles!

73. $0.500 \text{ g Ga} \times \dfrac{1 \text{ mol Ga}}{69.72 \text{ g Ga}} = 0.00717 \text{ mol Ga}$

$0.672 \text{ g Ga}_x\text{O}_y - 0.500 \text{ g Ga} = 0.172 \text{ g O}$

$0.172 \text{ g O} \times \dfrac{1 \text{ mol O}}{16.00 \text{ g O}} = 0.0108 \text{ mol O}$

$\dfrac{\text{Ga } 0.00717}{0.00717} \dfrac{\text{O } 0.0108}{0.00717} = \text{Ga}_{1.00}\text{O}_{1.51}$ The empirical formula is Ga_2O_3.

75. $1 \text{ molecule H}_2\text{O} \times \dfrac{1 \text{ mol H}_2\text{O}}{6.02 \times 10^{23} \text{ molecules H}_2\text{O}} \times \dfrac{18.02 \text{ g H}_2\text{O}}{1 \text{ mol H}_2\text{O}} \times \dfrac{1 \text{ cm}^3 \text{ H}_2\text{O}}{1.00 \text{ g H}_2\text{O}}$

$$= 2.99 \times 10^{-23} \text{ cm}^3 \text{ H}_2\text{O}$$

77. $\text{MM of C}_{12}\text{H}_{22}\text{O}_{11} = 12(12.01 \text{ g C}) + 22(1.01 \text{ g H}) + 11(16.00 \text{ g O})$
$= 144.12 \text{ g C} + 22.22 \text{ g H} + 176.00 \text{ g O}$
$= 342.34 \text{ g/mol}$

$1.00 \text{ g C}_{12}\text{H}_{22}\text{O}_{11} \times \dfrac{1 \text{ mol}}{342.34 \text{ g}} \times \dfrac{12 \text{ mol C}}{1 \text{ mol C}_{12}\text{H}_{22}\text{O}_{11}} \times \dfrac{6.02 \times 10^{23} \text{ atoms C}}{1 \text{ mol C}}$

$$= 2.11 \times 10^{22} \text{ atoms C}$$

79. $1 \text{ molecule vitamin K} \times \dfrac{1 \text{ mol vitamin K}}{6.02 \times 10^{23} \text{ molecules vitamin K}} \times \dfrac{173 \text{ g vitamin K}}{1 \text{ mol vitamin K}}$

$\times \dfrac{76.3 \text{ g C}}{100 \text{ g vitamin K}} \times \dfrac{1 \text{ mol C}}{12.01 \text{ g C}} \times \dfrac{6.02 \times 10^{23} \text{ atoms C}}{1 \text{ mol C}} = 11 \text{ atoms C}$

(*Note*: Since the number of atoms must be a whole number, the answer is 11 atoms not 11.0 atoms.)

81. Volume of Cu atom in cm^3:

$\dfrac{1 \text{ atom Cu}}{0.0118 \text{ nm}^3} \times \left(\dfrac{1 \times 10^9 \text{ nm}}{1 \text{ m}}\right)^3 \times \left(\dfrac{1 \text{ m}}{100 \text{ cm}}\right)^3 = \dfrac{8.47 \times 10^{22} \text{ atoms Cu}}{1 \text{ cm}^3}$

$\dfrac{8.47 \times 10^{22} \text{ atoms Cu}}{1 \text{ cm}^3} \times \dfrac{1 \text{ cm}^3}{8.92 \text{ g Cu}} \times \dfrac{63.55 \text{ g Cu}}{1 \text{ mol Cu}} = \dfrac{6.04 \times 10^{23} \text{ atoms Cu}}{\text{mol Cu}}$

Stoichiometry

Section 10.1 *Interpreting a Chemical Equation*

1.   General Equation: $2\,A + 3\,B \rightarrow C + 2\,D$

   (a) $2 \cancel{\text{moles A}} \times \dfrac{1 \text{ mole C}}{2 \cancel{\text{moles A}}} = 1 \text{ mole C}$

   (b) $2 \cancel{\text{liters A}} \times \dfrac{2 \text{ liters D}}{2 \cancel{\text{liters A}}} = 2 \text{ liters D}$

3.   General Equation: $A + 3\,B \rightarrow 2\,C$

   (a) $10.0 \text{ g A} + 15.0 \text{ g B} = 25.0 \text{ g C}$
   (b) $75.0 \text{ g C} - 50.0 \text{ g A} = 25.0 \text{ g B}$

5.   (a) $2\,KNO_3(s) \rightarrow 2\,KNO_2(s) + O_2(g)$
      MM of KNO_3 = 39.10 g K + 14.01 g N + $3(16.00$ g O$) = 101.11$ g/mol
      MM of KNO_2 = 39.10 g K + 14.01 g N + $2(16.00$ g O$) = 85.11$ g/mol
      MM of O_2 = $2(16.00$ g O$) = 32.0$ g/mol

$$2(101.11 \text{ g } KNO_3) \rightarrow 2(85.11 \text{ g } KNO_2) + 32.00 \text{ g } O_2$$
$$202.22 \text{ g } KNO_3 \rightarrow 170.22 \text{ g } KNO_2 + 32.00 \text{ g } O_2$$
$$202.22 \text{ g} \rightarrow 202.22 \text{ g}$$

   (b) $2\,Al(s) + 3\,Br_2(s) \rightarrow 2\,AlBr_3(s)$
      MM of Al   = 26.98 g/mol
      MM of Br_2 = $2(79.90$ g Br$) = 159.80$ g/mol
      MM of $AlBr_3$ = 26.98 g Al + $3(79.90$ g Br$) = 266.68$ g/mol

$$2(26.98 \text{ g Al}) + 3(159.80 \text{ g } Br_2) \rightarrow 2(266.68 \text{ g } AlBr_3)$$
$$53.96 \text{ g Al} + 479.40 \text{ g } Br_2 \rightarrow 533.36 \text{ g } AlBr_3$$
$$533.36 \text{ g} \rightarrow 533.36 \text{ g}$$

Section 10.2 *Mole–Mole Problems*

7. $2\,H_2(g) + O_2(g) \rightarrow 2\,H_2O(g)$

    Moles of O_2 that react:

    $$0.500 \;\cancel{\text{mol } H_2} \;\times\; \frac{1 \text{ mol } O_2}{2 \;\cancel{\text{mol } H_2}} \;=\; 0.250 \text{ mol } O_2 \text{ react}$$

    Moles of H_2O produced:

    $$0.500 \;\cancel{\text{mol } H_2} \;\times\; \frac{2 \text{ mol } H_2O}{2 \;\cancel{\text{mol } H_2}} \;=\; 0.500 \text{ mol } H_2O \text{ produced}$$

9. $2\,Fe(s) + 3\,Cl_2(g) \rightarrow 2\,FeCl_3(s)$

    Moles of Cl_2 that react:

    $$0.333 \;\cancel{\text{mol } Fe} \;\times\; \frac{3 \text{ mol } Cl_2}{2 \;\cancel{\text{mol } Fe}} \;=\; 0.500 \text{ mol } Cl_2 \text{ react}$$

    Moles of $FeCl_3$ produced:

    $$0.333 \;\cancel{\text{mol } Fe} \;\times\; \frac{2 \text{ mol } FeCl_3}{2 \;\cancel{\text{mol } Fe}} \;=\; 0.333 \text{ mol } FeCl_3 \text{ produced}$$

11. $C_3H_8(g) + 5\,O_2(g) \rightarrow 3\,CO_2(g) + 4\,H_2O(g)$

    Moles of $C_3H_8(g)$ that react:

    $$1.75 \;\cancel{\text{mol } O_2} \;\times\; \frac{1 \text{ mol } C_3H_8}{5 \;\cancel{\text{mol } O_2}} \;=\; 0.350 \text{ mol } C_3H_8 \text{ react}$$

    Moles of $CO_2(g)$ produced:

    $$1.75 \;\cancel{\text{mol } O_2} \;\times\; \frac{3 \text{ mol } CO_2}{5 \;\cancel{\text{mol } O_2}} \;=\; 1.05 \text{ mol } CO_2 \text{ produced}$$

Section 10.3 *Types of Stoichiometry Problems*

	Given	Unknown	Type of Stoichiometry
13.	mass	mass	mass–mass problem
15.	mass	volume	mass–volume problem
17.	mass	mass	mass–mass problem

Section 10.4 *Mass–Mass Problems*

19. $2\,Zn(s) + O_2(g) \rightarrow 2\,ZnO(s)$

   MM of Zn = 65.39 g/mol
   MM of ZnO = 81.39 g/mol

   $$2.36 \; \text{g Zn} \; \times \; \frac{1 \; \text{mol Zn}}{65.39 \; \text{g Zn}} \; \times \; \frac{2 \; \text{mol ZnO}}{2 \; \text{mol Zn}} \; \times \; \frac{81.39 \; \text{g ZnO}}{1 \; \text{mol ZnO}} \; = \; 2.94 \; \text{g ZnO}$$

21. $2\,Bi(s) + 3\,Cl_2(g) \rightarrow 2\,BiCl_3(s)$

   MM of Bi = 208.98 g/mol
   MM of BiCl$_3$ = 315.33 g/mol

   $$3.45 \; \text{g Bi} \; \times \; \frac{1 \; \text{mol Bi}}{208.98 \; \text{g Bi}} \; \times \; \frac{2 \; \text{mol BiCl}_3}{2 \; \text{mol Bi}} \; \times \; \frac{315.33 \; \text{g BiCl}_3}{1 \; \text{mol BiCl}_3} \; = \; 5.21 \; \text{g BiCl}_3$$

23. $Cu(s) + 2\,AgNO_3(aq) \rightarrow Cu(NO_3)_2(aq) + 2\,Ag(s)$

   MM of Cu = 63.55 g/mol
   MM of Ag = 107.87 g/mol

   $$0.615 \; \text{g Cu} \; \times \; \frac{1 \; \text{mol Cu}}{63.55 \; \text{g Cu}} \; \times \; \frac{2 \; \text{mol Ag}}{1 \; \text{mol Cu}} \; \times \; \frac{107.87 \; \text{g Ag}}{1 \; \text{mol Ag}} \; = \; 2.09 \; \text{g Ag}$$

25. $2\,Co(s) + 3\,HgCl_2(aq) \rightarrow 2\,CoCl_3(aq) + 3\,Hg(l)$

   MM of Co = 58.93 g/mol
   MM of Hg = 200.59 g/mol

   $$1.25 \; \text{g Co} \; \times \; \frac{1 \; \text{mol Co}}{58.93 \; \text{g Co}} \; \times \; \frac{3 \; \text{mol Hg}}{2 \; \text{mol Co}} \; \times \; \frac{200.59 \; \text{g Hg}}{1 \; \text{mol Hg}} \; = \; 6.38 \; \text{g Hg}$$

27. $2\,Na_3PO_4(aq) + 3\,Ca(OH)_2(aq) \rightarrow Ca_3(PO_4)_2(s) + 6\,NaOH(aq)$

   MM of Na$_3$PO$_4$ = 163.94 g/mol
   MM of Ca$_3$(PO$_4$)$_2$ = 310.18 g/mol

   $$1.78 \; \text{g Na}_3\text{PO}_4 \; \times \; \frac{1 \; \text{mol Na}_3\text{PO}_4}{163.94 \; \text{g Na}_3\text{PO}_4} \; \times \; \frac{1 \; \text{mol Ca}_3(\text{PO}_4)_2}{2 \; \text{mol Na}_3\text{PO}_4} \; \times \; \frac{310.18 \; \text{g Ca}_3(\text{PO}_4)_2}{1 \; \text{mol Ca}_3(\text{PO}_4)_2}$$

   $$= \; 1.68 \; \text{g Ca}_3(\text{PO}_4)_2$$

Section 10.5 *Mass–Volume Problems*

29. $Fe_2(CO_3)_3(s) \rightarrow Fe_2O_3(s) + 3\,CO_2(g)$

MM of $Fe_2(CO_3)_3$ = 291.73 g/mol

$$1.59\;\cancel{\text{g } Fe_2(CO_3)_3} \times \frac{1\;\cancel{\text{mol } Fe_2(CO_3)_3}}{291.73\;\cancel{\text{g } Fe_2(CO_3)_3}} \times \frac{3\;\cancel{\text{mol } CO_2}}{1\;\cancel{\text{mol } Fe_2(CO_3)_3}} \times \frac{22.4\;\cancel{\text{L } CO_2}}{1\;\cancel{\text{mol } CO_2}}$$

$$\times \frac{1000\;\text{mL } CO_2}{1\;\cancel{\text{L } CO_2}} = 366\;\text{mL } CO_2$$

31. $2\,LiHCO_3(s) \rightarrow Li_2CO_3(s) + H_2O(g) + CO_2(g)$

MM of $LiHCO_3$ = 67.96 g/mol

$$1.59\;\cancel{\text{g } LiHCO_3} \times \frac{1\;\cancel{\text{mol } LiHCO_3}}{67.96\;\cancel{\text{g } LiHCO_3}} \times \frac{1\;\cancel{\text{mol } CO_2}}{2\;\cancel{\text{mol } LiHCO_3}} \times \frac{22.4\;\cancel{\text{L } CO_2}}{1\;\cancel{\text{mol } CO_2}}$$

$$\times \frac{1000\;\text{mL } CO_2}{1\;\cancel{\text{L } CO_2}} = 262\;\text{mL } CO_2$$

33. $Mg(s) + H_2SO_4(aq) \rightarrow MgSO_4(aq) + H_2(g)$

MM of Mg = 24.31 g/mol

$$225\;\cancel{\text{mL } H_2} \times \frac{1\;\cancel{\text{L } H_2}}{1000\;\cancel{\text{mL } H_2}} \times \frac{1\;\cancel{\text{mol } H_2}}{22.4\;\cancel{\text{L } H_2}} \times \frac{1\;\cancel{\text{mol } Mg}}{1\;\cancel{\text{mol } H_2}} \times \frac{24.31\;\text{g } Mg}{1\;\cancel{\text{mol } Mg}}$$

$$= 0.244\;\text{g } Mg$$

35. $2\,H_2O_2(l) \rightarrow 2\,H_2O(l) + O_2(g)$

MM of H_2O_2 = 34.02 g/mol

$$55.0\;\cancel{\text{mL } O_2} \times \frac{1\;\cancel{\text{L } O_2}}{1000\;\cancel{\text{mL } O_2}} \times \frac{1\;\cancel{\text{mol } O_2}}{22.4\;\cancel{\text{L } O_2}} \times \frac{2\;\cancel{\text{mol } H_2O_2}}{1\;\cancel{\text{mol } O_2}} \times \frac{34.02\;\text{g } H_2O_2}{1\;\cancel{\text{mol } H_2O_2}}$$

$$= 0.167\;\text{g } H_2O_2$$

Section 10.6 Volume–Volume Problems

37. $2\,CO(g) + O_2(g) \rightarrow 2\,CO_2(g)$

$$2.00 \;\cancel{L\,CO} \times \frac{1\,L\,O_2}{2\;\cancel{L\,CO}} = 1.00\,L\,O_2$$

39. $H_2(g) + I_2(g) \rightarrow 2\,HI(g)$

$$125 \;\cancel{mL\,H_2} \times \frac{1\,mL\,I_2}{1\;\cancel{mL\,H_2}} = 125\,mL\,I_2$$

41. $3\,H_2(g) + N_2(g) \rightarrow 2\,NH_3(g)$

$$45.0 \;\cancel{mL\,NH_3} \times \frac{1\,mL\,N_2}{2\;\cancel{mL\,NH_3}} = 22.5\,mL\,N_2$$

43. $2\,Cl_2(g) + 3\,O_2(g) \rightarrow 2\,Cl_2O_3(g)$

$$1.75 \;\cancel{L\,Cl_2O_3} \times \frac{2\,L\,Cl_2}{2\;\cancel{L\,Cl_2O_3}} = 1.75\,L\,Cl_2$$

$$1.75 \;\cancel{L\,Cl_2} \times \frac{1000\,mL\,Cl_2}{1\;\cancel{L\,Cl_2}} = 1750\,mL\,Cl_2$$

45. $2\,SO_2(g) + O_2(g) \rightarrow 2\,SO_3(g)$

$$25.0 \;\cancel{L\,O_2} \times \frac{2\,L\,SO_3}{1\;\cancel{L\,O_2}} = 50.0\,L\,SO_3$$

47. $2\,N_2(g) + 5\,O_2(g) \rightarrow 2\,N_2O_5(g)$

$$500.0 \;\cancel{cm^3\,N_2O_5} \times \frac{2\,cm^3\,N_2}{2\;\cancel{cm^3\,N_2O_5}} = 500.0\,cm^3\,N_2$$

Section 10.7 *The Limiting Reactant Concept*

49. $N_2(g) + O_2(g) \rightarrow 2\,NO(g)$

$$1.00 \; \cancel{mol\,N_2} \times \frac{2 \; mol \; NO}{1 \; \cancel{mol\,N_2}} = 2.00 \; mol \; NO$$

$$1.50 \; \cancel{mol\,O_2} \times \frac{2 \; mol \; NO}{1 \; \cancel{mol\,O_2}} = 3.00 \; mol \; NO$$

Therefore, N_2 is the limiting reactant because it will be consumed before O_2. This reaction will produce 2.00 mol of NO.

51. $2\,NO(g) + O_2(g) \rightarrow 2\,NO_2(g)$

$$1.00 \; \cancel{mol\,NO} \times \frac{2 \; mol \; NO_2}{2 \; \cancel{mol\,NO}} = 1.00 \; mol \; NO_2$$

$$1.00 \; \cancel{mol\,O_2} \times \frac{2 \; mol \; NO_2}{1 \; \cancel{mol\,O_2}} = 2.00 \; mol \; NO_2$$

Therefore, NO is the limiting reactant because it will be consumed before O_2. This reaction will produce 1.00 mol of NO_2.

53. $2\,H_2(g) + O_2(g) \rightarrow 2\,H_2O(g)$

$$5.00 \; \cancel{mol\,H_2} \times \frac{2 \; mol \; H_2O}{2 \; \cancel{mol\,H_2}} = 5.00 \; mol \; H_2O$$

$$5.00 \; \cancel{mol\,O_2} \times \frac{2 \; mol \; H_2O}{1 \; \cancel{mol\,O_2}} = 10.0 \; mol \; H_2O$$

Therefore, H_2 is the limiting reactant because it will be consumed before O_2. This reaction will produce 5.00 mol of H_2O.

55. $2\,C_2H_6(g) + 7\,O_2(g) \rightarrow 4\,CO_2(g) + 6\,H_2O(g)$

$$1.00 \; \cancel{mol\,C_2H_6} \times \frac{6 \; mol \; H_2O}{2 \; \cancel{mol\,C_2H_6}} = 3.00 \; mol \; H_2O$$

$$5.00 \; \cancel{mol\,O_2} \times \frac{6 \; mol \; H_2O}{7 \; \cancel{mol\,O_2}} = 4.29 \; mol \; H_2O$$

Therefore, C_2H_6 is the limiting reactant because it will be consumed before O_2. This reaction will produce 3.00 mol of H_2O.

57. $Co(s) + S(s) \rightarrow CoS(s)$

(1) $1.50 \text{ mol Co} \times \dfrac{1 \text{ mol CoS}}{1 \text{ mol Co}} = 1.50 \text{ mol CoS}$

$2.00 \text{ mol S} \times \dfrac{1 \text{ mol CoS}}{1 \text{ mol S}} = 2.00 \text{ mol CoS}$

Therefore, the limiting reactant is Co because it will be consumed before S. This reaction will produce 1.50 mol of CoS.

After the reaction:
mol of Co = 1.50 − 1.50 = 0.00 mol
mol of S = 2.00 − 1.50 = 0.50 mol
mol of CoS = 0.00 + 1.50 = 1.50 mol

(2) $3.00 \text{ mol Co} \times \dfrac{1 \text{ mol CoS}}{1 \text{ mol Co}} = 3.00 \text{ mol CoS}$

$2.00 \text{ mol S} \times \dfrac{1 \text{ mol CoS}}{1 \text{ mol S}} = 2.00 \text{ mol CoS}$

Therefore, the limiting reactant is S because it will be consumed before Co. This reaction will produce 2.00 mol of CoS.

After the reaction:
mol of Co = 3.00 − 2.00 = 1.00 mol
mol of S = 2.00 − 2.00 = 0.00 mol
mol of CoS = 0.00 + 2.00 = 2.00 mol

Section 10.8 *Limiting Reactant Problems*

59. $FeO(l) + Mg(l) \rightarrow Fe(l) + MgO(s)$

$40.0 \text{ g FeO} \times \dfrac{1 \text{ mol FeO}}{71.85 \text{ g FeO}} \times \dfrac{1 \text{ mol Fe}}{1 \text{ mol FeO}} \times \dfrac{55.85 \text{ g Fe}}{1 \text{ mol Fe}} = 31.1 \text{ g Fe}$

$10.0 \text{ g Mg} \times \dfrac{1 \text{ mol Mg}}{24.31 \text{ g Mg}} \times \dfrac{1 \text{ mol Fe}}{1 \text{ mol Mg}} \times \dfrac{55.85 \text{ g Fe}}{1 \text{ mol Fe}} = 23.0 \text{ g Fe}$

Therefore, Mg is the limiting reactant because it will be consumed before FeO. This reaction will produce 23.0 g Fe.

61. $Fe_2O_3(l) + 2\,Al(l) \rightarrow 2\,Fe(l) + Al_2O_3(g)$

$$175\ \cancel{g\ Fe_2O_3} \times \frac{1\ \cancel{mol\ Fe_2O_3}}{159.70\ \cancel{g\ Fe_2O_3}} \times \frac{2\ \cancel{mol\ Fe}}{1\ \cancel{mol\ Fe_2O_3}} \times \frac{55.85\ g\ Fe}{\cancel{mol\ Fe}} = 122\ g\ Fe$$

$$37.5\ \cancel{g\ Al} \times \frac{1\ \cancel{mol\ Al}}{26.98\ \cancel{g\ Al}} \times \frac{2\ \cancel{mol\ Fe}}{2\ \cancel{mol\ Al}} \times \frac{55.85\ g\ Fe}{1\ \cancel{mol\ Fe}} = 77.6\ g\ Fe$$

Therefore, Al is the limiting reactant because it will be consumed before Fe_2O_3. This reaction will produce 77.6 g Fe.

63. $Mg(OH)_2(s) + H_2SO_4(l) \rightarrow MgSO_4(s) + 2\,H_2O(l)$

$$1.00\ \cancel{g\ Mg(OH)_2}(s) \times \frac{1\ \cancel{mol\ Mg(OH)_2}}{58.33\ \cancel{g\ Mg(OH)_2}} \times \frac{1\ \cancel{mol\ MgSO_4}}{1\ \cancel{mol\ Mg(OH)_2}} \times \frac{120.38\ g\ MgSO_4}{1\ \cancel{mol\ MgSO_4}}$$

$$= 2.06\ g\ MgSO_4$$

$$0.605\ \cancel{g\ H_2SO_4} \times \frac{1\ \cancel{mol\ H_2SO_4}}{98.09\ \cancel{g\ H_2SO_4}} \times \frac{1\ \cancel{mol\ MgSO_4}}{1\ \cancel{mol\ H_2SO_4}} \times \frac{120.38\ g\ MgSO_4}{1\ \cancel{mol\ MgSO_4}}$$

$$= 0.742\ g\ MgSO_4$$

Therefore, H_2SO_4 is the limiting reactant because it will be consumed before $Mg(OH)_2$. This reaction will produce 0.742 g $MgSO_4$.

65. $2\,Al(OH)_3(s) + 3\,H_2SO_4(l) \rightarrow Al_2(SO_4)_3(aq) + 6\,H_2O(l)$

$$1.00\ \cancel{g\ Al(OH)_3} \times \frac{1\ \cancel{mol\ Al(OH)_3}}{78.01\ \cancel{g\ Al(OH)_3}} \times \frac{6\ \cancel{mol\ H_2O}}{2\ \cancel{mol\ Al(OH)_3}} \times \frac{18.02\ g\ H_2O}{1\ \cancel{mol\ H_2O}}$$

$$= 0.693\ g\ H_2O$$

$$3.00\ \cancel{g\ H_2SO_4} \times \frac{1.00\ \cancel{mol\ H_2SO_4}}{98.09\ \cancel{g\ H_2SO_4}} \times \frac{6\ \cancel{mol\ H_2O}}{3\ \cancel{mol\ H_2SO_4}} \times \frac{18.02\ g\ H_2O}{1\ \cancel{mol\ H_2O}}$$

$$= 1.10\ g\ H_2O$$

Therefore, $Al(OH)_3$ is the limiting reactant because it will be consumed before H_2SO_4. This reaction will produce 0.693 g H_2O.

67. $N_2(g) + 2 O_2(g) \rightarrow 2 NO_2(g)$

$$45.0 \text{ mL } N_2 \times \frac{2 \text{ mL } NO_2}{1 \text{ mL } N_2} = 90.0 \text{ mL } NO_2$$

$$95.0 \text{ mL } O_2 \times \frac{2 \text{ mL } NO_2}{2 \text{ mL } O_2} = 95.0 \text{ mL } NO_2$$

Therefore, N_2 is the limiting reactant because it will be consumed before O_2. This reaction will produce 90.0 mL NO_2.

69. $2 N_2(g) + 3 O_2(g) \rightarrow 2 N_2O_3(g)$

$$70.0 \text{ mL } N_2 \times \frac{2 \text{ mL } N_2O_3}{2 \text{ mL } N_2} = 70.0 \text{ mL } N_2O_3$$

$$45.0 \text{ mL } O_2 \times \frac{2 \text{ mL } N_2O_3}{3 \text{ mL } O_2} = 30.0 \text{ mL } N_2O_3$$

Therefore, O_2 is the limiting reactant because it will be consumed before N_2. This reaction will produce 30.0 mL N_2O_3.

71. $2 SO_2(g) + O_2(g) \rightarrow 2 SO_3(g)$

$$3.00 \text{ L } SO_2 \times \frac{2 \text{ L } SO_3}{2 \text{ L } SO_2} = 3.00 \text{ L } SO_3$$

$$1.25 \text{ L } O_2 \times \frac{2 \text{ L } SO_3}{1 \text{ L } O_2} = 2.50 \text{ L } SO_3$$

Therefore, O_2 is the limiting reactant because it will be consumed before SO_2. This reaction will produce 2.50 L SO_3.

73. $4 HCl(g) + O_2(g) \rightarrow 2 Cl_2(g) + 2 H_2O(g)$

$$50.0 \text{ L } HCl \times \frac{2 \text{ L } Cl_2}{4 \text{ L } HCl} = 25.0 \text{ L } Cl_2$$

$$10.0 \text{ L } O_2 \times \frac{2 \text{ L } Cl_2}{1 \text{ L } O_2} = 20.0 \text{ L } Cl_2$$

Therefore, O_2 is the limiting reactant because it will be consumed before HCl. This reaction will produce 20.0 L Cl_2.

Section 10.9 *Percent Yield*

75. Actual yield: 10.4 g acetone; theoretical yield: 11.6 g acetone

$$\text{Percent yield:} \quad \frac{10.4 \text{ g}}{11.6 \text{ g}} \times 100 = 89.7\%$$

77. Actual yield: 1.29 g $NaNO_2$; theoretical yield: 1.22 g $NaNO_2$

$$\text{Percent yield:} \quad \frac{1.29 \text{ g}}{1.22 \text{ g}} \times 100 = 106\%$$

General Exercises

79. The units associated with molar mass are grams per mole (g/mol).

81. $(NH_4)_2Cr_2O_7(s) \rightarrow Cr_2O_3(s) + 4\,H_2O(l) + N_2(g)$

MM of $(NH_4)_2Cr_2O_7$ = 252.10 g/mol
MM of Cr_2O_3 = 152.00 g/mol

$$1.54 \text{ g } (NH_4)_2Cr_2O_7 \times \frac{1 \text{ mol } (NH_4)_2Cr_2O_7}{252.10 \text{ g } (NH_4)_2Cr_2O_7} \times \frac{1 \text{ mol } Cr_2O_3}{1 \text{ mol } (NH_4)_2Cr_2O_7}$$

$$\times \frac{152.00 \text{ g } Cr_2O_3}{1 \text{ mol } Cr_2O_3} = 0.929 \text{ g } Cr_2O_3$$

83. $3\,MnO_2(s) + 4\,Al(s) \rightarrow 3\,Mn(s) + 2\,Al_2O_3(s)$

MM of Mn = 54.94 g/mol
MM of Al = 26.98 g/mol

$$1.00 \text{ kg Mn} \times \frac{1000 \text{ g Mn}}{1 \text{ kg Mn}} \times \frac{1 \text{ mol Mn}}{54.94 \text{ g Mn}} \times \frac{4 \text{ mol Al}}{3 \text{ mol Mn}} \times \frac{26.98 \text{ g Al}}{1 \text{ mol Al}}$$

$$= 655 \text{ g Al}$$

85. $Sb_2S_3(s) + 6\,HCl(aq) \rightarrow 2\,SbCl_3(aq) + 3\,H_2S(g)$

MM of Sb_2S_3 = 339.71 g/mol

$$3.00\ \cancel{g\ Sb_2S_3} \times \frac{1\ \cancel{mol\ Sb_2S_3}}{339.71\ \cancel{g\ Sb_2S_3}} \times \frac{3\ \cancel{mol\ H_2S}}{1\ \cancel{mol\ Sb_2S_3}} \times \frac{22.4\ L\ H_2S}{1\ \cancel{mol\ H_2S}} = 0.593\ L\ H_2S$$

87. $2\,H_2O(l) \rightarrow 2\,H_2(g) + O_2(g)$

MM of H_2O = 18.02 g/mol

$$100.0\ \cancel{mL\ H_2O} \times \frac{1.00\ \cancel{g\ H_2O}}{1\ \cancel{mL\ H_2O}} \times \frac{1\ \cancel{mol\ H_2O}}{18.02\ \cancel{g\ H_2O}} \times \frac{2\ \cancel{mol\ H_2}}{2\ \cancel{mol\ H_2O}} \times \frac{22.4\ \cancel{L\ H_2}}{1\ \cancel{mol\ H_2}}$$

$$= 124\ L\ H_2$$

89. $C_3H_8(g) + 5\,O_2(g) \rightarrow 3\,CO_2(g) + 4\,H_2O(g)$

MM of C_3H_8 = 44.11 g/mol
MM of H_2O = 18.02 g/mol

$$10.0\ \cancel{g\ C_3H_8} \times \frac{1\ \cancel{mol\ C_3H_8}}{44.11\ \cancel{g\ C_3H_8}} \times \frac{4\ \cancel{mol\ H_2O}}{1\ \cancel{mol\ C_3H_8}} \times \frac{18.02\ g\ H_2O}{1\ \cancel{mol\ H_2O}} = 16.3\ g\ H_2O$$

91. The limiting reactant is gasoline; the excess reactant is oxygen.

The Gaseous State

Section 11.1 *Properties of Gases*

1.   The observed properties of a gas are as follows:
      (1) Gases have an indefinite shape.
      (2) Gases may be expanded.
      (3) Gases may be compressed.
      (4) Gases have low densities.
      (5) Gases mix spontaneously to form homogeneous mixtures.

Section 11.2 *Atmospheric Pressure*

3. | Units | Standard Pressure |
      |---|---|
      | (a) atmospheres | 1 atm |
      | (b) millimeters of mercury | 760 mm Hg |
      | (c) torr | 760 torr |
      | (d) centimeters of mercury | 76 cm Hg |

5.   (a) $5.25 \text{ atm} \times \dfrac{760 \text{ mm Hg}}{1 \text{ atm}} = 3990 \text{ mm Hg}$

      (b) $5.25 \text{ atm} \times \dfrac{760 \text{ torr}}{1 \text{ atm}} = 3990 \text{ torr}$

      (c) $5.25 \text{ atm} \times \dfrac{76 \text{ cm Hg}}{1 \text{ atm}} = 399 \text{ cm Hg}$

      (d) $5.25 \text{ atm} \times \dfrac{29.9 \text{ in. Hg}}{1 \text{ atm}} = 157 \text{ in. Hg}$

7. (a) $28.8 \; \text{in. Hg} \times \dfrac{1 \text{ atm}}{29.9 \text{ in. Hg}} = 0.963 \text{ atm}$

(b) $28.8 \; \text{in. Hg} \times \dfrac{760 \text{ mm Hg}}{29.9 \text{ in. Hg}} = 732 \text{ mm Hg}$

(c) $28.8 \; \text{in. Hg} \times \dfrac{76 \text{ cm Hg}}{29.9 \text{ in. Hg}} = 73.2 \text{ cm Hg}$

(d) $28.8 \; \text{in. Hg} \times \dfrac{760 \text{ torr}}{29.9 \text{ in. Hg}} = 732 \text{ torr}$

Section 11.3 *Variables Affecting Gas Pressure*

9. The three variables that affect the pressure of a gas are *volume, temperature,* and *number of molecules.*

11.

Change	Observation	Explanation
(a) volume increases	pressure decreases	molecules are farther apart and collide less frequently
(b) temperature increases	pressure increases	molecules are moving faster and collide with higher frequency and more energy
(c) moles of gas increase	pressure increases	more molecules have more collisions

13.

Change	Observation
(a) volume increases	pressure decreases
(b) temperature increases	pressure increases
(c) moles of gas decrease	pressure decreases

Section 11.4 *Boyle's Law*

15. Pressure vs. Volume

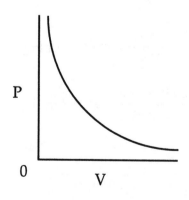

17. $P_1 \times V_{factor} = P_2$

$$0.750 \text{ atm} \times \frac{250.0 \, \cancel{mL}}{655.0 \, \cancel{mL}} = 0.286 \text{ atm}$$

19. $P_1 \times V_{factor} = P_2$

$$15.0 \text{ psi} \times \frac{50.0 \, \cancel{mL}}{44.0 \, \cancel{mL}} = 17.0 \text{ psi}$$

Section 11.5 *Charles' Law*

21. <u>Volume vs. Kelvin Temperature</u>

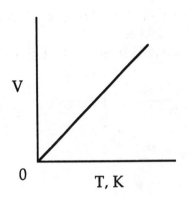

23. $V_1 \times T_{factor} = V_2$

$25°C + 273 = 298 \text{ K}$
$50°C + 273 = 323 \text{ K}$

$$335 \text{ mL} \times \frac{323 \, \cancel{K}}{298 \, \cancel{K}} = 363 \text{ mL O}_2$$

25. $V_1 \times T_{factor} = V_2$

$0°C + 273 = 273 \text{ K}$
$100°C + 273 = 373 \text{ K}$

$$80.0 \text{ cm}^3 \times \frac{373 \, \cancel{K}}{273 \, \cancel{K}} = 109 \text{ cm}^3 \, F_2$$

Section 11.6 *Gay–Lussac's Law*

27. <u>Pressure vs. Kelvin Temperature</u>

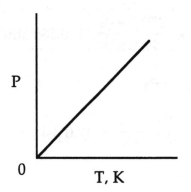

29. $P_1 \times T_{factor} = P_2$

$20°C + 273 = 293 \text{ K}$
$200°C + 273 = 473 \text{ K}$

$760 \text{ torr} \times \dfrac{473 \ \cancel{K}}{293 \ \cancel{K}} = 1230 \text{ torr}$

31. $P_1 \times T_{factor} = P_2$

$0°C + 273 = 273 \text{ K}$
$100°C + 273 = 373 \text{ K}$

$76.0 \text{ cm Hg} \times \dfrac{373 \ \cancel{K}}{273 \ \cancel{K}} = 104 \text{ cm Hg}$

Section 11.7 *Combined Gas Law*

33.

	P	V	T
initial	772 mm Hg	100.0 mL	21°C + 273 = 294 K
final	760 mm Hg	V_2	273 K

$V_1 \quad \times \quad P_{factor} \quad \times \quad T_{factor} \quad = \quad V_2$

$100.0 \text{ mL} \times \dfrac{772 \ \cancel{\text{mm Hg}}}{760 \ \cancel{\text{mm Hg}}} \times \dfrac{273 \ \cancel{K}}{294 \ \cancel{K}} = 94.3 \text{ mL H}_2$

35.

	P	V	T
initial	760 torr	2.00 L	273 K
final	365 torr	V_2	75°C + 273 = 348 K

$$V_1 \quad \times \quad P_{factor} \quad \times \quad T_{factor} \quad = \quad V_2$$

$$2.00 \text{ L} \times \frac{760 \text{ torr}}{365 \text{ torr}} \times \frac{348 \text{ K}}{273 \text{ K}} = 5.31 \text{ L air}$$

37.

	P	V	T
initial	760 torr	1250 mL	273 K
final	P_2	255 mL	300°C + 273 = 573 K

$$P_1 \quad \times \quad V_{factor} \quad \times \quad T_{factor} \quad = \quad P_2$$

$$760 \text{ torr} \times \frac{1250 \text{ mL}}{255 \text{ mL}} \times \frac{573 \text{ K}}{273 \text{ K}} = 7820 \text{ torr}$$

39.

	P	V	T
initial	225 mm Hg	500.0 mL	−125°C + 273 = 148 K
final	P_2	220.0 mL	100°C + 273 = 373 K

$$P_1 \quad \times \quad V_{factor} \quad \times \quad T_{factor} \quad = \quad P_2$$

$$225 \text{ mm Hg} \times \frac{500.0 \text{ mL}}{220.0 \text{ mL}} \times \frac{373 \text{ K}}{148 \text{ K}} = 1290 \text{ mm Hg}$$

41.

	P	V	T
initial	760 torr	50.0 mL	273 K
final	350 torr	350.0 mL	T_2

$$T_1 \quad \times \quad P_{factor} \quad \times \quad V_{factor} \quad = \quad T_2$$

$$273 \text{ K} \times \frac{350 \text{ torr}}{760 \text{ torr}} \times \frac{350.0 \text{ mL}}{50.0 \text{ mL}} = 880 \text{ K}$$

$$880 \text{ K} - 273 = 607°C$$

Section 11.8 *The Vapor Pressure Concept*

43. The general relationship between the vapor pressure of a liquid and its temperature is as follows: *As the temperature of a liquid increases, so does its vapor pressure.*

45. <u>Temperature</u>   <u>Vapor Pressure</u>
     (a) 25°C   23.8 mm Hg
     (b) 50°C   92.5 mm Hg

Section 11.9 *Dalton's Law*

47. $P_{nitrogen} = 587$ mm Hg
    $P_{oxygen} = 158$ mm Hg
    $P_{argon} = 7$ mm Hg

    $P_{nitrogen} + P_{oxygen} + P_{argon} = P_{atmosphere}$
    587 mm Hg + 158 mm Hg + 7 mm Hg = 752 mm Hg

49. $P_{total} = 1$ atm = 760 mm Hg
    $P_{sulfur\ dioxide} = 150$ mm Hg
    $P_{sulfur\ trioxide} = 475$ mm Hg

    $P_{sulfur\ dioxide} + P_{sulfur\ trioxide} + P_{oxygen} = P_{total}$
    $P_{oxygen} = P_{total} - P_{sulfur\ dioxide} - P_{sulfur\ trioxide}$
    $P_{oxygen} = 760$ mm Hg $- 150$ mm Hg $- 475$ mm Hg $= 135$ mm Hg

51. A gas may be collected over water to determine its volume. The volume of gas collected displaces an equal volume of water (see Figures 3.3 and 11.11). When collecting a gas over water to determine volume, we must assume that the gas is not very soluble in water.

53. $P_{total} = 766$ torr
    $P_{water\ vapor} = 17.5$ mm Hg = 17.5 torr

    $P_{water\ vapor} + P_{oxygen} = P_{total}$
    $P_{oxygen} = P_{total} - P_{water\ vapor}$
    $P_{oxygen} = 766$ torr $- 17.5$ torr $= 748.5$ torr

Section 11.10 *Ideal Gas Behavior*

55. The characteristics of an ideal gas are:
    (1) Gases are mostly empty space and are composed of tiny molecules, which have a negligible volume.
    (2) Gas molecules demonstrate rapid motion, move in straight lines, and travel in random directions.
    (3) Gas molecules show no attraction for one another.
    (4) Gas molecules have elastic collisions.
    (5) The average kinetic energy of gas molecules is proportional to the Kelvin temperature.

57. A real gas behaves most like an ideal gas under the conditions of *high temperature* and *low pressure*.

59.

Description	Gas
(a) highest kinetic energy	all gases are equal
(b) lowest kinetic energy	all gases are equal
(c) highest velocity	He atoms
(d) lowest velocity	Ar atoms

(*Note:* Since all three gases are at the same temperature, each has the same kinetic energy.)

61. An ideal gas at 0 K exerts no pressure (0 mm Hg).

Section 11.11 *Ideal Gas Law*

63. $P = \dfrac{n\,R\,T}{V}$

$n = 0.500 \text{ mol } H_2$

$T = 25°C + 273 = 298 \text{ K}$

$V = 50.0 \text{ mL} \times \dfrac{1 \text{ L}}{1000 \text{ mL}} = 0.0500 \text{ L}$

$P = \dfrac{0.500 \text{ mol} \times 298 \text{ K}}{0.0500 \text{ L}} \times \dfrac{0.0821 \text{ atm} \cdot \text{L}}{1 \text{ mol} \cdot \text{K}} = 245 \text{ atm}$

65. $n = \dfrac{P\,V}{R\,T}$

$V = 10.0 \text{ L}$

$T = 373 \text{ K}$

$P = 125 \text{ psi} \times \dfrac{1 \text{ atm}}{14.7 \text{ psi}} = 8.50 \text{ atm}$

$n = \dfrac{8.50 \text{ atm} \times 10.0 \text{ L}}{373 \text{ K}} \times \dfrac{1 \text{ mol} \cdot \text{K}}{0.0821 \text{ atm} \cdot \text{L}} = 2.78 \text{ mol } N_2O$

67. $MM = \dfrac{g\,R\,T}{P\,V}$

$\quad g = 2.14\text{ g}$

$\quad V = 1.00\text{ L}$

$\quad P = 1\text{ atm}$

$\quad T = 273\text{ K}$

$$MM = \frac{2.14\text{ g} \times 273\text{ K}}{1\text{ atm} \times 1.00\text{ L}} \times \frac{0.0821\text{ atm} \cdot \text{L}}{1\text{ mol} \cdot \text{K}} = 48.0\text{ g/mol}$$

69. $MM = \dfrac{g\,R\,T}{P\,V}$

$\quad g = 1.95\text{ g}$

$\quad V = 3.00\text{ L}$

$\quad P = 1.25\text{ atm}$

$\quad T = 20°C + 273 = 293\text{ K}$

$$MM = \frac{1.95\text{ g} \times 293\text{ K}}{1.25\text{ atm} \times 3.00\text{ L}} \times \frac{0.0821\text{ atm} \cdot \text{L}}{1\text{ mol} \cdot \text{K}} = 12.5\text{ g/mol}$$

71. $n = \dfrac{P\,V}{R\,T}$

$\quad V = 1550\text{ mL} \times \dfrac{1\text{ L}}{1000\text{ mL}} = 1.55\text{ L}$

$\quad P = 0.945\text{ atm}$

$\quad T = 50°C + 273 = 323\text{ K}$

$$n = \frac{0.945\text{ atm} \times 1.55\text{ L}}{323\text{ K}} \times \frac{1\text{ mol} \cdot \text{K}}{0.0821\text{ atm} \cdot \text{L}} = 0.0552\text{ mol Cl}_2$$

$$0.0552\text{ mol Cl}_2 \times \frac{70.90\text{ g Cl}_2}{1\text{ mol Cl}_2} = 3.92\text{ g Cl}_2$$

General Exercises

73. $2500\text{ in.}^2 \times \dfrac{14.7\text{ lb}}{1\text{ in.}^2} = 37{,}000\text{ lb}$

75. $P_{nitrogen} = 0.79\text{ atm} \times \dfrac{14.7\text{ psi}}{1\text{ atm}} = 12\text{ psi}$

$\quad P_{oxygen} = 3.07\text{ psi}$

$\quad P_{argon} = 7.55\text{ torr} \times \dfrac{1\text{ atm}}{760\text{ torr}} \times \dfrac{14.7\text{ psi}}{1\text{ atm}} = 0.146\text{ psi}$

$P_{nitrogen} + P_{oxygen} + P_{argon} = P_{total}$

$12\text{ psi} + 3.07\text{ psi} + 0.146\text{ psi} = 15\text{ psi}$

77. P_{total} = 764 mm Hg

$P_{water\ vapor}$ = 19.8 mm Hg

$P_{oxygen} + P_{water\ vapor} = P_{total}$

$P_{oxygen} = P_{total} - P_{water\ vapor}$

P_{oxygen} = 764 mm Hg − 19.8 mm Hg = 744.2 mm Hg

	P	V	T
initial	744.2 mm Hg	42.5 mL	22°C + 273 = 295 K
final	760 mm Hg	V_2	273 K

$$V_1 \times P_{factor} \times T_{factor} = V_2$$

$$42.5\ mL \times \frac{744.2\ \text{mm Hg}}{760\ \text{mm Hg}} \times \frac{273\ \text{K}}{295\ \text{K}} = 38.5\ mL\ O_2$$

79. $P = \dfrac{nRT}{V}$

$n = 1.51 \times 10^{24}$ atoms Kr $\times \dfrac{1\ mol\ Kr}{6.02 \times 10^{23}\ \text{atoms Kr}} = 2.51$ mol Kr

$T = 25°C + 273 = 298\ K$

$V = 5.00\ L$

$$P = \frac{2.51\ \text{mol} \times 298\ \text{K}}{5.00\ \text{L}} \times \frac{0.0821\ atm \cdot \text{L}}{1\ \text{mol} \cdot \text{K}} = 12.3\ atm$$

81. $V = \dfrac{nRT}{P}$

$n = 3.38 \times 10^{22}$ molecules NO $\times \dfrac{1\ mol\ NO}{6.02 \times 10^{23}\ \text{molecules NO}} = 0.0561$ mol NO

$T = 100°C + 273 = 373\ K$

$P = 255\ \text{torr} \times \dfrac{1\ atm}{760\ \text{torr}} = 0.336\ atm$

$$V = \frac{0.0561\ \text{mol} \times 373\ \text{K}}{0.336\ \text{atm}} \times \frac{0.0821\ \text{atm} \cdot L}{1\ \text{mol} \cdot \text{K}} = 5.11\ L\ NO$$

83. $$\frac{0.0821 \; \cancel{atm} \cdot L}{1 \; mol \cdot K} \times \frac{760 \; torr}{1 \; \cancel{atm}} = \frac{62.4 \; torr \cdot L}{mol \cdot K}$$

85. Deep-sea divers breathe a mixture that is mostly helium gas. Since helium atoms are lighter than nitrogen molecules, the helium gas moves faster over the vocal cords and produces a higher-pitched voice.

Chemical Bonding

Section 12.1 *The Chemical Bond Concept*

1. A metal atom loses valence electrons to become a positively charged cation; a nonmetal atom gains valence electrons to become a negatively charged anion. The attraction between the resulting cation and anion is called an ionic bond.

3.
Element	Before Ionic Bond	After Ionic Bond
Mg	2 valence e$^-$	0 valence e$^-$
S	6 valence e$^-$	8 valence e$^-$

5.
	Compound	Bond		Compound	Bond
(a)	H_2O	covalent	(b)	NaCl	ionic
(c)	CH_4	covalent	(d)	ZnO	ionic

7.
	Substance	Representative Unit
(a)	C_2H_5OH	molecule
(b)	SrI_2	formula unit
(c)	CO	molecule
(d)	$CaCO_3$	formula unit

9.
	Substance	Representative Unit
(a)	Ne	atom
(b)	F_2	molecule
(c)	CF_2Cl_2	molecule
(d)	UF_6	formula unit

Section 12.2 *Ionic Bonds*

11.
	Ion	Ionic Charge		Ion	Ionic Charge
(a)	Na ion	1+ (Group 1)	(b)	Mg ion	2+ (Group 2)
(c)	Sn ion	4+ (Group 14)	(d)	Al ion	3+ (Group 13)

13.

	Ion	Ionic Charge		Ion	Ionic Charge
(a)	F ion	1– (Group 17)	(b)	Br ion	1– (Group 17)
(c)	S ion	2– (Group 16)	(d)	N ion	3– (Group 15)

15.

	Ion	Electron Configuration
(a)	Li^+	$1s^2$
(b)	Al^{3+}	$1s^2\,2s^2\,2p^6$
(c)	Ca^{2+}	$1s^2\,2s^2\,2p^6\,3s^2\,3p^6$
(d)	Mg^{2+}	$1s^2\,2s^2\,2p^6$

17.

	Ion	Electron Configuration
(a)	Cl^-	$1s^2\,2s^2\,2p^6\,3s^2\,3p^6$
(b)	I^-	$1s^2\,2s^2\,2p^6\,3s^2\,3p^6\,4s^2\,3d^{10}\,4p^6\,5s^2\,4d^{10}\,5p^6$
(c)	S^{2-}	$1s^2\,2s^2\,2p^6\,3s^2\,3p^6$
(d)	P^{3-}	$1s^2\,2s^2\,2p^6\,3s^2\,3p^6$

19.

	Ion	Isoelectronic Noble Gas
(a)	Li^+	He
(b)	K^+	Ar
(c)	Ca^{2+}	Ar
(d)	Ra^{2+}	Rn

21.

	Ion	Isoelectronic Noble Gas
(a)	Li^+	He
(b)	Al^{3+}	Ne
(c)	Ca^{2+}	Ar
(d)	Mg^{2+}	Ne

23.

	Ion	Isoelectronic Noble Gas
(a)	Cl^-	Ar
(b)	I^-	Xe
(c)	S^{2-}	Ar
(d)	P^{3-}	Ar

25.
(a) Li atomic radius > Li ion radius
(b) Mg atomic radius > Mg ion radius
(c) F atomic radius < F ion radius
(d) O atomic radius < O ion radius

27. The only true statement is (a). The corrected statements are:
(b) The ionic radius of a metal atom is *less than* its atomic radius.
(c) The ionic radius of a nonmetal atom is *greater than* its atomic radius.
(d) The simplest representative particle is a *formula unit.*

Section 12.3 *Covalent Bonds*

29. (a) The sum of the atomic radii is *greater than* the bond length in H—I.
    (b) The sum of the atomic radii is *greater than* the bond length in N—O.

31. All of the statements are false. The corrected statements are:
    (a) A covalent bond is formed by *sharing a pair of electrons between two nonmetal atoms.*
    (b) The bonding electrons are *delocalized over both atoms* in the covalent bond.
    (c) The bond length in a covalent bond is *less than* the sum of the two atomic radii.
    (d) An amount of energy equal to the bond energy *is required to break* a covalent bond.

Section 12.4 *Electron Dot Formulas of Molecules*

33.

Molecule	Valence Electrons	Electron Dot	Structural Formula
(a) H_2	$1 + 1 = 2\ e^-$	H:H	H—H
(b) F_2	$7 + 7 = 14\ e^-$	:F̈:F̈:	F — F
(c) HBr	$1 + 7 = 8\ e^-$	H:B̈r:	H— Br
(d) NH_3	$5 + 3(1) = 8\ e-$	H:N̈:H with H above	H—N—H with H above

35.

Molecule	Valence Electrons	Electron Dot	Structural Formula
(a) HONO	$1 + 2(6) + 5 = 18\ e^-$	H:Ö:N::Ö	H—O—N=O
(b) SO_2	$6 + 2(6) = 18\ e^-$	:Ö:S̈::Ö	O—S=O
(c) C_2H_4	$2(4) + 4(1) = 12\ e^-$	H:C::C:H with H H below	H—C=C—H with H H below
(d) C_2H_2	$2(4) + 2(1) = 10\ e^-$	H:C:::C:H	H—C≡C—H

37.

Molecule	Valence Electrons	Electron Dot	Structural Formula
(a) CH_4	$4 + 4(1) = 8\ e^-$	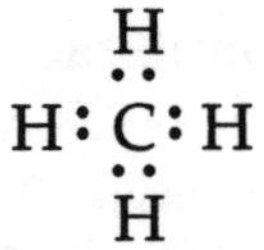	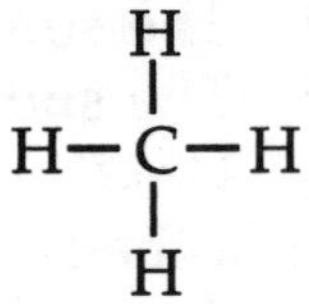
(b) OF_2	$6 + 2(7) = 20\ e^-$	 $:\!\ddot{F}\!:\!\ddot{O}\!:\!\ddot{F}\!:$	F—O—F
(c) H_2O_2	$2(1) + 2(6) = 14\ e^-$	 $H\!:\!\ddot{O}\!:\!\ddot{O}\!:\!H$	H—O—O—H
(d) NF_3	$5 + 3(7) = 26\ e^-$		

Section 12.5 *Electron Dot Formulas of Polyatomic Ions*

39.

Polyatomic Ion	Valence Electrons	Electron Dot	Structural Formula
(a) BrO^-	$7 + 6 + 1 = 14\ e^-$	 $:\!\ddot{B}r\!:\!\ddot{O}\!:\ ^-$	$[\,Br\!-\!O\,]^-$
(b) BrO_2^-	$7 + 2(6) + 1 = 20\ e^-$	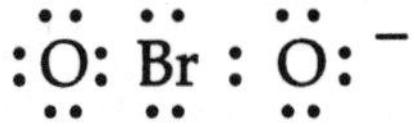 $:\!\ddot{O}\!:\!\ddot{B}r\!:\!\ddot{O}\!:\ ^-$	$[\,O\!-\!Br\!-\!O\,]^-$
(c) BrO_3^-	$7 + 3(6) + 1 = 26\ e^-$	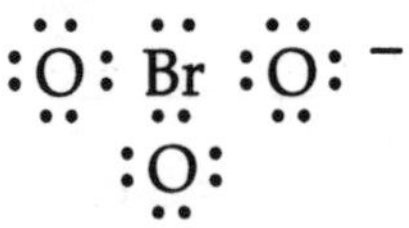	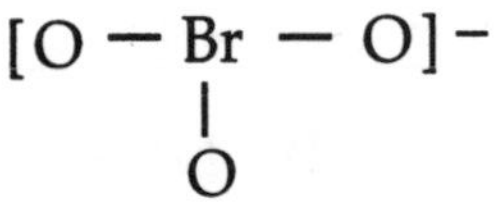
(d) BrO_4^-	$7 + 4(6) + 1 = 32\ e^-$	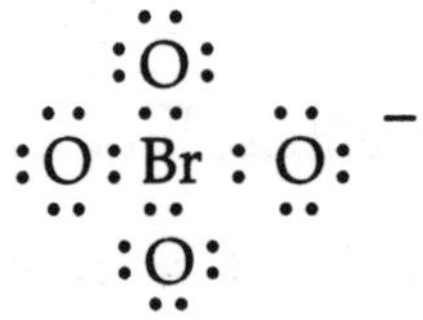	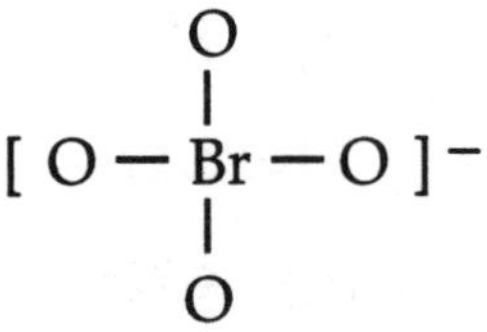

41.

Polyatomic Ion	Valence Electrons	Electron Dot	Structural Formula
(a) SO_4^{2-}	$6 + 4(6) + 2 = 32\ e^-$	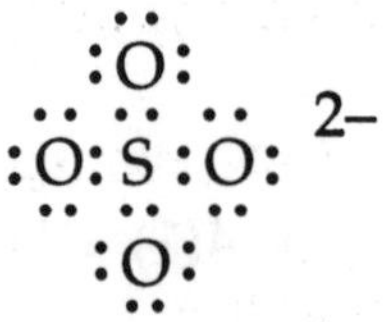	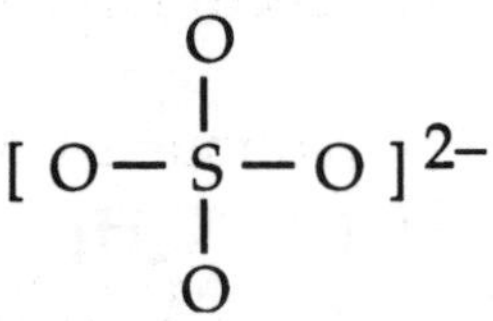

(b) HSO_4^- $1 + 6 + 4(6) + 1 = 32\ e^-$

$$\left[H-O-\underset{\underset{O}{|}}{\overset{\overset{O}{||}}{S}}-O \right]^-$$

(c) SO_3^{2-} $6 + 3(6) + 2 = 26\ e^-$

$$\left[O-\underset{\underset{O}{|}}{S}-O \right]^{2-}$$

(d) HSO_3^- $1 + 6 + 3(6) + 1 = 26\ e^-$

$$\left[H-O-\underset{\underset{O}{|}}{S}-O \right]^-$$

43.

Polyatomic Ion	Valence Electrons	Electron Dot	Structural Formula
(a) H_3O^+	$3(1) + 6 - 1 = 8\ e^-$		$\left[H-\underset{\underset{H}{|}}{O}-H \right]^+$
(b) OH^-	$6 + 1 + 1 = 8\ e^-$		$[O-H]^-$
(c) HS^-	$1 + 6 + 1 = 8\ e^-$		$[H-S]^-$
(d) CN^-	$4 + 5 + 1 = 10\ e^-$		$[C \equiv N]^-$

Section 12.6 *Polar Covalent Bonds*

45. Electronegativity down a group in the periodic table generally decreases.

47. Nonmetals are more electronegative than metals.

49.
More Electronegative		More Electronegative	
(a)	**Cl** > Br	(b)	**O** > S
(c)	**Se** > As	(d)	N < **F**

(*Note:* The more electronegative element is in bold.)

51.

	Bond	Polarity		Bond	Polarity
(a)	Br—Cl	$3.0 - 2.8 = 0.2$	(b)	Br—F	$4.0 - 2.8 = 1.2$
(c)	I—Cl	$3.0 - 2.5 = 0.5$	(d)	I—Br	$2.8 - 2.5 = 0.3$

53. Polar Bonds Using Delta Notation

(a) δ^+ H—S δ^-

(b) δ^- O—S δ^+

(c) δ^+ N—F δ^-

(d) δ^+ S—Cl δ^-

(*Note:* δ^- indicates the more electronegative atom and δ^+ indicates the more electropositive atom.)

Section 12.7 *Nonpolar Covalent Bonds*

55.

	Bond	Polarity	Classification
(a)	Cl—Cl	$3.0 - 3.0 = 0$	nonpolar
(b)	Cl—N	$3.0 - 3.0 = 0$	nonpolar
(c)	N—H	$3.0 - 2.1 = 0.9$	polar
(d)	H—P	$2.1 - 2.1 = 0$	nonpolar

Thus, (a), (b), and (d) are nonpolar.

57. H_2, N_2, and O_2 occur naturally as diatomic molecules.

Section 12.8 *Coordinate Covalent Bonds*

(*Note*: Coordinate covalent bonds are indicated by a dash, —.)

59.

Molecule	Valence Electrons	Electron Dot	Coord. Cov. Bond
HBrO	$1 + 7 + 6 = 14\ e^-$	H:Br̈:Ö:	H:Br̈ — Ö:

61.

Molecule	Valence Electrons	Electron Dot	Coord. Cov. Bond
HBrO₂	$1 + 7 + 2(6) = 20\ e^-$	H:Br̈:Ö: :Ö:	H:Br̈ — Ö: | :Ö:

63.

Polyatomic Ion	Valence Electrons	Electron Dot	Coord. Cov. Bond
NH₄⁺	$5 + 4(1) - 1 = 8\ e^-$	H H:N̈:H ⁺ H	H H:N̈:H ⁺ | H

65.

Polyatomic Ion	Valence Electrons	Electron Dot	Coord. Cov. Bond

NO_3^- $5 + 3(6) + 1 = 24\ e^-$

Section 12.9 *Shapes of Molecules*

67.

	Formula	Electron Pair	Molecular Shape	Bond Angle
(a)	SiH_4	tetrahedral	tetrahedral	109.5°
(b)	PH_3	tetrahedral	trigonal (pyramid)	107°

69.

	Formula	Electron Pair	Molecular Shape	Bond Angle
(a)	HBr	tetrahedral	linear	—
(b)	Br_2O	tetrahedral	angular (bent)	104.5°

71. Each C—F bond in a CF_4 molecule is polar, but the symmetrical arrangement of the four bonds (that is, tetrahedral) produces a nonpolar molecule.

General Exercises

73.

	Substance	Representative Unit
(a)	Cr	atom
(b)	P_4	molecule
(c)	CrP	formula unit

75.

	Ions	Chemical Formula
(a)	Ca^{2+} and $2\ I^-$	CaI_2
(b)	Ra^{2+} and O^{2-}	RaO
(c)	Ga^{3+} and $3\ F^-$	GaF_3
(d)	$3\ Ba^{2+}$ and $2\ P^{3-}$	Ba_3P_2

77.

	Ions	Chemical Formula
(a)	$2\ Al^{3+}$ and $3\ CO_3^{2-}$	$Al_2(CO_3)_3$
(b)	Sr^{2+} and $2\ OH^-$	$Sr(OH)_2$
(c)	$3\ Ag^+$ and PO_4^{3-}	Ag_3PO_4
(d)	Cd^{2+} and $2\ NO_3^-$	$Cd(NO_3)_2$

79. The radius of a sodium ion is less than the radius of a sodium atom because the sodium ion has one less energy level ($3s$). In addition, the ion has one more proton than electron, which draws the electrons closer to the nucleus.

81.

Bond	Polarity
B—Cl	$3.0 - 2.0 = 1.0$ (polar)

83. | Bond | Polarity |
| --- | --- |
| H—P | $2.1 - 2.1 = 0$ (nonpolar) |

85. | Polar Bond and Delta Notation | δ^+ Ge—Cl δ^- |

87. | Molecule | Valence Electrons | Electron Dot | Structural Formula |
| --- | --- | --- | --- |
| SiH_4 | $4 + 4(1) = 8\ e^-$ | H:Si:H with H above and H below | H—Si—H with H above and H below |

89. | Polyatomic Ion | Valence Electrons | Electron Dot | Structural Formula |
| --- | --- | --- | --- |
| AsO_3^{3-} | $5 + 3(6) + 3 = 26\ e^-$ | :Ö:As:Ö: $^{3-}$ with :Ö: below | $[O—As—O]^{3-}$ with O below |

91. In a molecule, the valence electrons are free to move about the entire region of all bonded atoms. Thus, the valence electrons are said to be delocalized.

93. | Molecule | Structural Formula #1 | Structural Formula #2 |
| --- | --- | --- |
| SO_2 | $O = S - O$ | $O - S = O$ |

95. | Molecule | Valence Electrons | Electron Dot | Structural Formula |
| --- | --- | --- | --- |
| BF_3 | $3 + 3(7) = 24\ e^-$ | :F:B:F: with :F: below | F—B—F with F below |

97. | Molecule | Valence Electrons | Electron Dot | Structural Formula |
| --- | --- | --- | --- |
| XeO_2 | $8 + 2(6) = 20\ e^-$ | :Ö:Xe:Ö: | $O—Xe—O$ |

99. The electron pair geometry in both H_2O and H_3O^+ is tetrahedral. However, in H_3O^+ the molecular shape is tetrahedral and in H_2O the molecular shape is angular (bent).

Liquids and Solids

Section 13.1 *Properties of Liquids*

1.   General Properties of Liquids
   (1) Liquids have an indefinite shape but a fixed volume.
   (2) Liquids usually flow readily.
   (3) Liquids do not expand or compress to any degree.
   (4) Liquids have a high density compared to gases.
   (5) Liquids that are soluble mix uniformly.

3.

	Substance	Temperature	Physical State
(a)	H_2O	$-20.0°C$	solid
(b)	H_2O	$120.0°C$	gas
(c)	NH_3	$-195.0°C$	solid
(d)	NH_3	$0.0°C$	gas
(e)	$CHCl_3$	$-55.5°C$	liquid
(f)	$CHCl_3$	$100.0°C$	gas

Section 13.2 *Vapor Pressure, Viscosity, Surface Tension*

5.   Water molecules in the liquid state evaporate to the gaseous state and form water vapor. Simultaneously, some water molecules in the vapor condense back to the liquid. By definition, the vapor pressure of water is the total pressure exerted by the water molecules in the vapor when the rate of evaporation is equal to the rate of condensation.

7.   The viscosity of a liquid is a measure of its resistance to flow. Since the intermolecular attraction in water is strong, water has less tendency to flow than other liquids with molecules the same size.

9. If the molecules in a liquid have a strong intermolecular attraction:
   (a) the vapor pressure is low.   (b) the boiling point is *high*.
   (c) the viscosity is *high*.   (d) the surface tension is *high*.

11. The relationship between the vapor pressure of a liquid and its temperature is: *As the temperature of a liquid increases, the vapor pressure* increases.

13.
	Temperature	Vapor Pressure
(a)	15°C	~70 mm Hg
(b)	30°C	~300 mm Hg

15. The temperature at which the vapor pressure is equal to the atmospheric pressure is the boiling point. Thus, the boiling point of acetone is 56°C because the vapor pressure at this temperature is 760 mm Hg.

Section 13.3 *The Intermolecular Bond Concept*

17.
	Liquid	Intermolecular Attraction
(a)	C_8H_{18}	dispersion forces
(b)	CH_3–OH	dispersion forces, hydrogen bonds
(c)	CH_3–Cl	dispersion forces, dipole forces
(d)	CH_3–O–CH_3	dispersion forces, dipole forces

19.
	Liquid	Higher Vapor Pressure
(a)	CH_3COOH or C_2H_5Cl	C_2H_5Cl (weaker attraction)
(b)	C_2H_5OH or CH_3OCH_3	CH_3OCH_3 (weaker attraction)

21.
	Liquid	Higher Viscosity
(a)	CH_3COOH or C_2H_5Cl	CH_3COOH (stronger attraction)
(b)	C_2H_5OH or CH_3OCH_3	C_2H_5OH (stronger attraction)

Section 13.4 *Properties of Solids*

23. General Properties of Solids
   (1) Solids have a definite shape and a fixed volume.
   (2) Solids are either crystalline or noncrystalline.
   (3) Solids do not compress or expand to any degree.
   (4) Solids usually have a slightly higher density than their corresponding liquids.
   (5) Solids do not mix by diffusion.

25.

	Substance	Temperature	Physical State
(a)	Ga	0°C	solid
(b)	Ga	100°C	liquid
(c)	Sn	0°C	solid
(d)	Sn	100°C	solid
(e)	Hg	0°C	liquid
(f)	Hg	100°C	liquid

Section 13.5 *Crystalline Solids*

27. Three examples of crystalline solids are NaCl (ionic solid), Cl_2 (molecular solid), and Na (metallic solid).

29.

	Crystalline Solid	Type of Particles
(a)	ionic solid	ions
(b)	molecular solid	molecules
(c)	metallic solid	metal atoms

31.

	Crystalline Solid	Classification
(a)	Zn	metallic solid
(b)	ZnO	ionic solid
(c)	P_4	molecular solid
(d)	IBr	molecular solid

Section 13.6 *Changes of Physical State*

33.

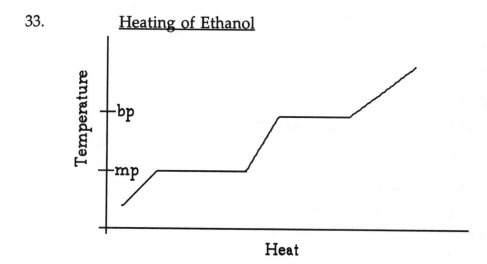

Heating of Ethanol

35. Energy required to melt ice:

$$125 \; \cancel{g} \; \times \; \frac{80.0 \text{ cal}}{1 \; \cancel{g}} \; = \; 1.00 \times 10^4 \text{ cal} \;\; (10.0 \text{ kcal})$$

37. Energy required to heat water:

$$25.0 \; \cancel{g} \; \times \; \frac{1.00 \text{ cal}}{1 \; \cancel{g} \times \cancel{°C}} \; \times \; (100.0 - 25.0) \cancel{°C} \; = \; 1880 \text{ cal} \;\; (1.88 \text{ kcal})$$

Energy required to vaporize water:

$$25.0 \; \cancel{g} \; \times \; \frac{540 \text{ cal}}{1 \; \cancel{g}} \; = \; 13{,}500 \text{ cal} \;\; (13.5 \text{ kcal})$$

Total heat energy required:

$$1{,}880 \text{ cal} + 13{,}500 \text{ cal} \; = \; 15{,}400 \text{ cal} \;\; (15.4 \text{ kcal})$$

39. Energy required to melt ice:

$$115 \; \cancel{g} \; \times \; \frac{80.0 \text{ cal}}{1 \; \cancel{g}} \; = \; 9200 \text{ cal} \;\; (9.20 \text{ kcal})$$

Energy required to heat water:

$$115 \; \cancel{g} \; \times \; \frac{1.00 \text{ cal}}{1 \; \cancel{g} \times \cancel{°C}} \; \times \; (100.0 - 0.0) \cancel{°C} \; = \; 11{,}500 \text{ cal} \;\; (11.5 \text{ kcal})$$

Energy required to vaporize water:

$$115 \; \cancel{g} \; \times \; \frac{540 \text{ cal}}{1 \; \cancel{g}} \; = \; 62{,}100 \text{ cal} \;\; (62.1 \text{ kcal})$$

Total heat energy required:

$$9{,}200 \text{ cal} + 11{,}500 \text{ cal} + 62{,}100 \text{ cal} \; = \; 82{,}800 \text{ cal} \;\; (82.8 \text{ kcal})$$

41. Energy required to heat ice:

$$38.5 \; \cancel{g} \; \times \; \frac{0.50 \text{ cal}}{1 \; \cancel{g} \times \cancel{°C}} \; \times \; [0.0 - (-20.0)] \cancel{°C} \; = \; 385 \text{ cal} \;\; (0.385 \text{ kcal})$$

Energy required to melt ice:

$$38.5 \; \cancel{g} \; \times \; \frac{80.0 \text{ cal}}{1 \; \cancel{g}} \; = \; 3080 \text{ cal} \;\; (3.08 \text{ kcal})$$

Energy required to heat water:

$$38.5 \, \cancel{g} \times \frac{1.00 \text{ cal}}{1 \, \cancel{g} \times \cancel{^\circ C}} \times (100.0 - 0.0) \cancel{^\circ C} = 3850 \text{ cal} \ (3.85 \text{ kcal})$$

Energy required to vaporize water:

$$38.5 \, \cancel{g} \times \frac{540 \text{ cal}}{1 \, \cancel{g}} = 20{,}800 \text{ cal} \ (20.8 \text{ kcal})$$

Total heat energy required:

385 cal + 3080 cal + 3850 cal + 20,800 cal = 28,100 cal (28.1 kcal)

43. Energy required to heat ice:

$$100.0 \, \cancel{g} \times \frac{0.50 \text{ cal}}{1 \, \cancel{g} \times \cancel{^\circ C}} \times [0.0 - (-40.0)] \cancel{^\circ C} = 2{,}000 \text{ cal} \ (2.00 \text{ kcal})$$

Energy required to melt ice:

$$100.0 \, \cancel{g} \times \frac{80.0 \text{ cal}}{1 \, \cancel{g}} = 8000 \text{ cal} \ (8.00 \text{ kcal})$$

Energy required to heat water:

$$100.0 \, \cancel{g} \times \frac{1.00 \text{ cal}}{1 \, \cancel{g} \times \cancel{^\circ C}} \times (100.0 - 0.0) = 10{,}000 \text{ cal} \ (10.0 \text{ kcal})$$

Energy required to vaporize water:

$$100.0 \, \cancel{g} \times \frac{540 \text{ cal}}{1 \, \cancel{g}} = 54{,}000 \text{ cal} \ (54.0 \text{ kcal})$$

Energy required to heat steam:

$$100.0 \, \cancel{g} \times \frac{0.48 \text{ cal}}{1 \, \cancel{g} \times \cancel{^\circ C}} \times (125.0 - 100.0) \cancel{^\circ C} = 1200 \text{ cal} \ (1.20 \text{ kcal})$$

Total heat energy required:

2,000 cal + 8000 cal + 10,000 cal + 54,000 cal + 1200 cal

$$= 75{,}200 \text{ cal} \ (75.2 \text{ kcal})$$

Section 13.7 *Structure of Water*

45. In a water molecule, there are *two pairs* of bonding electrons and *two pairs* of nonbonding electrons.

47. The observed bond angle in a water molecule is 104.5°.

49. Structural Formula of Water with Delta Notation

51. Hydrogen Bonding in Hydrogen Fluoride

$$H{-}F \cdots\cdots\cdots H{-}F$$
$$\uparrow$$
$$H\ bond$$

Section 13.8 *Physical Properties of Water*

53. An "ammonia ice cube" *floats* because the density of solid ammonia is less than the density of liquid ammonia.

55. | Liquid | Higher Melting Point |
(a) H_2O or H_2S — H_2O (hydrogen bond attraction)
(b) H_2S or H_2Se — H_2Se (larger size, more attraction)

57. | Liquid | Higher Heat of Fusion |
(a) H_2O or H_2S — H_2O (hydrogen bond attraction)
(b) H_2S or H_2Se — H_2Se (larger size, more attraction)

59. | Property | As Molar Mass Increases |
(a) melting point — increases
(b) boiling point — increases
(c) heat of fusion — increases
(d) heat of vaporization — increases

Section 13.9 *Chemical Properties of Water*

61. $2 H_2O(l) \rightarrow 2 H_2(g) + O_2(g)$

63.
(a) $2 Li(s) + 2 H_2O(l) \rightarrow 2 LiOH(aq) + H_2(g)$
(b) $Na_2O(s) + H_2O(l) \rightarrow 2 NaOH(aq)$
(c) $CO_2(g) + H_2O(l) \rightarrow H_2CO_3(aq)$

65.
(a) $Ba(s) + 2 H_2O(l) \rightarrow Ba(OH)_2(aq) + H_2(g)$
(b) $N_2O_3(g) + H_2O(l) \rightarrow 2 HNO_2(aq)$
(c) $CaO(s) + H_2O(l) \rightarrow Ca(OH)_2(aq)$

67.
(a) $2 C_3H_6(g) + 9 O_2(g) \rightarrow 6 CO_2(g) + 6 H_2O(g)$
(b) $Na_2Cr_2O_7 \cdot 2 H_2O(s) \rightarrow Na_2Cr_2O_7(s) + 2 H_2O(g)$
(c) $2 HF(aq) + Ca(OH)_2(aq) \rightarrow CaF_2(aq) + 2 HOH(l)$

69.
(a) $2 C_4H_{10}(g) + 13 O_2(g) \rightarrow 8 CO_2(g) + 10 H_2O(g)$
(b) $Co(C_2H_3O_2)_2 \cdot 4 H_2O(s) \rightarrow Co(C_2H_3O_2)_2(s) + 4 H_2O(g)$
(c) $2 HNO_3(aq) + Ba(OH)_2(aq) \rightarrow Ba(NO_3)_2(aq) + 2 HOH(l)$

Section 13.10 *Hydrates*

71.

	Chemical Formula	Systematic Name
(a)	$MgSO_4 \cdot 7 H_2O$	magnesium sulfate heptahydrate
(b)	$Co(CN)_3 \cdot 3 H_2O$	cobalt(III) cyanide trihydrate
(c)	$MnSO_4 \cdot H_2O$	manganese(II) sulfate monohydrate
(d)	$Na_2Cr_2O_7 \cdot 2 H_2O$	sodium dichromate dihydrate

73.

	Systematic Name	Chemical Formula
(a)	sodium acetate trihydrate	$NaC_2H_3O_2 \cdot 3 H_2O$
(b)	calcium sulfate dihydrate	$CaSO_4 \cdot 2 H_2O$
(c)	potassium chromate tetrahydrate	$K_2CrO_4 \cdot 4 H_2O$
(d)	zinc sulfate heptahydrate	$ZnSO_4 \cdot 7 H_2O$

75.
(a) Percentage of water in $SrCl_2 \cdot 6 H_2O$

MM of $SrCl_2 = 87.62$ g $+ 2(35.45$ g$) = 158.52$ g

Percentage of water: $\dfrac{6(18.02 \text{ g})}{158.52 \text{ g} + 6(18.02 \text{ g})} \times 100 = 40.55\% \text{ } H_2O$

(b) Percentage of water in $K_2Cr_2O_7 \cdot 2 \, H_2O$

MM of $K_2Cr_2O_7 = 2(39.10 \, g) + 2(52.00 \, g) + 7(16.00 \, g) = 294.20 \, g$

Percentage of water: $\dfrac{2(18.02 \, g)}{294.20 \, g + 2(18.02 \, g)} \times 100 = 10.91\% \, H_2O$

(c) Percentage of water in $Co(CN)_3 \cdot 3 \, H_2O$

MM of $Co(CN)_3 = 58.93 \, g + 3(12.01 \, g) + 3(14.01 \, g) = 136.99 \, g$

Percentage of water: $\dfrac{3(18.02 \, g)}{136.99 \, g + 3(18.02 \, g)} \times 100 = 28.28\% \, H_2O$

(d) Percentage of water in $Na_2CrO_4 \cdot 4 \, H_2O$

MM of $Na_2CrO_4 = 2(22.99 \, g) + 52.00 \, g + 4(16.00 \, g) = 161.98 \, g$

Percentage of water: $\dfrac{4(18.02 \, g)}{161.98 \, g + 4(18.02 \, g)} \times 100 = 30.80\% \, H_2O$

77. (a) $NiCl_2 \cdot X \, H_2O(s) \rightarrow NiCl_2(s) + X \, H_2O(g)$

$21.7 \, \cancel{g \, H_2O} \times \dfrac{1 \, mol \, H_2O}{18.02 \, \cancel{g \, H_2O}} = 1.20 \, mol \, H_2O$

$78.3 \, \cancel{g \, NiCl_2} \times \dfrac{1 \, mol \, NiCl_2}{129.59 \, \cancel{g \, NiCl_2}} = 0.604 \, mol \, NiCl_2$

$NiCl_2 \cdot \dfrac{1.20}{0.604} \, H_2O \qquad \dfrac{1.20}{0.604} = 1.99 \approx 2$

Chemical Formula: $NiCl_2 \cdot 2 \, H_2O$

(b) $Sr(NO_3)_2 \cdot X \, H_2O(s) \rightarrow Sr(NO_3)_2(s) + X \, H_2O(g)$

$33.8 \, \cancel{g \, H_2O} \times \dfrac{1 \, mol \, H_2O}{18.02 \, \cancel{g \, H_2O}} = 1.88 \, mol \, H_2O$

$66.2 \, \cancel{g \, Sr(NO_3)_2} \times \dfrac{1 \, mol \, Sr(NO_3)_2}{211.64 \, \cancel{g \, Sr(NO_3)_2}} = 0.313 \, mol \, Sr(NO_3)_2$

$$Sr(NO_3)_2 \cdot \frac{1.88}{0.313} \; H_2O \qquad \frac{1.88}{0.313} = 6.01 \approx 6$$

Chemical Formula: $Sr(NO_3)_2 \cdot 6 \; H_2O$

(c) $CrI_3 \cdot X \; H_2O(s) \qquad \rightarrow \qquad CrI_3(s) + X \; H_2O(g)$

$$27.2 \; \cancel{g \; H_2O} \; \times \; \frac{1 \; mol \; H_2O}{18.02 \; \cancel{g \; H_2O}} = 1.51 \; mol \; H_2O$$

$$72.8 \; \cancel{g \; CrI_3} \; \times \; \frac{1 \; mol \; CrI_3}{432.70 \; \cancel{g \; CrI_3}} = 0.168 \; mol \; CrI_3$$

$$CrI_3 \cdot \frac{1.51}{0.168} \; H_2O \qquad \frac{1.51}{0.168} = 8.99 \approx 9$$

Chemical Formula: $CrI_3 \cdot 9 \; H_2O$

(d) $Ca(NO_3)_2 \cdot X \; H_2O(s) \rightarrow Ca(NO_3)_2(s) + X \; H_2O(g)$

$$30.5 \; \cancel{g \; H_2O} \; \times \; \frac{1 \; mol \; H_2O}{18.02 \; \cancel{g \; H_2O}} = 1.69 \; mol \; H_2O$$

$$69.5 \; \cancel{g \; Ca(NO_3)_2} \; \times \; \frac{1 \; mol \; Ca(NO_3)_2}{164.10 \; \cancel{g \; Ca(NO_3)_2}} = 0.424 \; mol \; Ca(NO_3)_2$$

$$Ca(NO_3)_2 \cdot \frac{1.69}{0.424} \; H_2O \qquad \frac{1.69}{0.424} = 3.99 \approx 4$$

Chemical Formula: $Ca(NO_3)_2 \cdot 4 \; H_2O$

General Exercises

79. Water covers about three-fourths (~75%) of Earth's surface.

81. The vapor pressure of water is 650 mm Hg at ~95°C. Therefore, the boiling point of water is ~95°C at 650 torr.

83. Sulfuric acid, H_2SO_4, is polar so we can predict that it behaves somewhat like water. That is, owing to surface tension sulfuric acid "raindrops" on Venus will form spheres similar to raindrops on Earth.

85. Energy released when ethylene glycol cools:

$$1250 \; \cancel{g} \times \frac{0.561 \; cal}{1 \; \cancel{g} \times \cancel{°C}} \times [25.0 - (-11.5)] \cancel{°C} = 25{,}600 \; cal \; (25.6 \; kcal)$$

Energy released when ethylene glycol solidifies:

$$1250 \; \cancel{g} \times \frac{43.3 \; cal}{1 \; \cancel{g}} = 54{,}100 \; cal \; (54.1 \; kcal)$$

Total heat energy released:

$25{,}600 \; cal + 54{,}100 \; cal = 79{,}700 \; cal \; (79.7 \; kcal)$

Solutions

Section 14.1 *Gases in Solution*

1. | | Change | Result |
   |---|---|---|
   | (a) | temperature of solution decreases | solubility of NH_3 gas *increases* |
   | (b) | partial pressure of NH_3 increases | solubility of NH_3 gas *increases* |

3. standard solubility $\times$ pressure factor $=$ new solubility

$$\frac{1.45\ g\ CO_2}{1\ L\ champagne} \times \frac{10.0\ \cancel{atm}}{1.00\ \cancel{atm}} = 14.5\ g\ CO_2/1\ L\ champagne$$

5. $\dfrac{0.63\ g\ Cl_2}{100\ g\ H_2O} \times \dfrac{1200\ \cancel{mm\ Hg}}{760\ \cancel{mm\ Hg}} = 0.99\ g\ Cl_2/100\ g\ H_2O$

Section 14.2 *Liquids in Solution*

7. (a) polar solute + polar solvent $=$ miscible
   (b) polar solute + nonpolar solvent $=$ immiscible

9. | | Solvent | Classification |
   |---|---|---|
   | (a) | H_2O | polar |
   | (b) | C_6H_{14} | nonpolar |
   | (c) | C_3H_6O | polar |
   | (d) | $CHCl_3$ | nonpolar |

11. | | Solvent | Miscible or Immiscible |
    |---|---|---|
    | (a) | C_7H_{16} | immiscible with H_2O |
    | (b) | CH_3OH | miscible with H_2O |
    | (c) | C_4H_8O | miscible with H_2O |
    | (d) | C_7H_8 | immiscible with H_2O |

13. Add several drops of the unknown liquid into a test tube containing water. If the unknown liquid is polar, it will be miscible with water. If the liquid is nonpolar, it will be immiscible and form two distinct layers in the test tube.

Section 14.3 *Solids in Solution*

15. (a) polar solute + polar solvent = soluble
    (b) nonpolar solute + polar solvent = insoluble
    (c) ionic solute + polar solvent = soluble

17.

Compound	Soluble or Insoluble
(a) naphthalene, $C_{10}H_8$	insoluble in H_2O
(b) potassium hydroxide, KOH	soluble in H_2O
(c) calcium acetate, $Ca(C_2H_3O_2)_2$	soluble in H_2O
(d) trichlorotoluene, $C_7H_5Cl_3$	insoluble in H_2O
(e) glycine, $C_2H_5NO_2$	soluble in H_2O
(f) lactic acid, $HC_3H_5O_3$	soluble in H_2O

19.

Vitamin	Soluble or Insoluble
(a) vitamin B_1, $C_{12}H_{18}Cl_2N_4OS$	water soluble
(b) vitamin B_3, $C_6H_6N_2O$	water soluble
(c) vitamin B_6, $C_8H_{11}NO_3$	water soluble
(d) vitamin C, $C_6H_8O_6$	water soluble
(e) vitamin D, $C_{27}H_{44}O$	fat soluble
(f) vitamin K, $C_{31}H_{46}O_2$	fat soluble

Section 14.4 *The Dissolving Process*

21. Fructose molecule, $C_6H_{12}O_6$, dissolved in water:

$$H_2O$$
$$\vdots$$
$$H_2O \cdots C_6H_{12}O_6 \cdots H_2O$$
$$\vdots$$
$$H_2O$$

23. (a) Lithium bromide, LiBr, dissolved in water:

(b) Calcium chloride, $CaCl_2$, dissolved in water:

Section 14.5 *Rate of Dissolving*

25. The three factors that increase the rate of dissolving of a solid solute in a liquid solvent are:
(1) heating the solution
(2) stirring the solution
(3) grinding the solid solute

Section 14.6 *Solubility and Temperature*

27. (a) ~ 35 g $NaCl$/100 g H_2O (b) ~ 35 g KCl/100 g H_2O

29. (a) ~ 36 g $NaCl$/100 g H_2O (b) ~ 41 g KCl/100 g H_2O

31. (a) ~ 0°C   (b) ~ 70°C

33. (a) ~ 100°C   (b) ~ 35°C

Section 14.7 *Unsaturated, Saturated, Supersaturated*

35. (a) supersaturated at 50°C
(b) saturated at 70°C
(c) unsaturated at 90°C

37. (a) supersaturated at 20°C
(b) saturated at 50°C
(c) unsaturated at 70°C

39. $25.0 \text{ g } H_2O \times \dfrac{40.0 \text{ g rock salt}}{100 \text{ g } H_2O} = 10.0 \text{ g rock salt}$

The solution is *saturated* at 30°C.

41. (a) ~ 80 g solute remains in solution
(b) ~ 20 g solute (100 g – 80 g) crystallizes from solution

Section 14.8 *Mass Percent Concentration*

43. $\dfrac{\text{mass of solute}}{\text{mass of solution}} \times 100 = m/m\ \%$

(a) $\dfrac{1.25\ \text{g NaCl}}{100.0\ \text{g solution}} \times 100 = 1.25\%$

(b) $\dfrac{2.50\ \text{g K}_2\text{Cr}_2\text{O}_7}{95.0\ \text{g solution}} \times 100 = 2.63\%$

(c) $\dfrac{10.0\ \text{g CaCl}_2}{250.0\ \text{g solution}} \times 100 = 4.00\%$

(d) $\dfrac{65.0\ \text{g sugar}}{125.0\ \text{g solution}} \times 100 = 52.0\%$

45. $\quad$ <u>Solution</u> $\qquad\qquad\qquad\qquad$ <u>Unit Factors</u>

(a) 1.50% KBr $\qquad \dfrac{1.50\ \text{g KBr}}{100.00\ \text{g solution}}$ and $\dfrac{100.00\ \text{g solution}}{1.50\ \text{g KBr}}$

$\dfrac{98.50\ \text{g H}_2\text{O}}{100.00\ \text{g solution}}$ and $\dfrac{100.00\ \text{g solution}}{98.50\ \text{g H}_2\text{O}}$

$\dfrac{1.50\ \text{g KBr}}{98.50\ \text{g H}_2\text{O}}$ and $\dfrac{98.50\ \text{g H}_2\text{O}}{1.50\ \text{g KBr}}$

(b) 2.50% AlCl$_3$ $\qquad \dfrac{2.50\ \text{g AlCl}_3}{100.00\ \text{g solution}}$ and $\dfrac{100.00\ \text{g solution}}{2.50\ \text{g AlCl}_3}$

$\dfrac{97.50\ \text{g H}_2\text{O}}{100.00\ \text{g solution}}$ and $\dfrac{100.00\ \text{g solution}}{97.50\ \text{g H}_2\text{O}}$

$\dfrac{2.50\ \text{g AlCl}_3}{97.50\ \text{g H}_2\text{O}}$ and $\dfrac{97.50\ \text{g H}_2\text{O}}{2.50\ \text{g AlCl}_3}$

(c) 3.75% AgNO$_3$ $\qquad \dfrac{3.75\ \text{g AgNO}_3}{100.00\ \text{g solution}}$ and $\dfrac{100.00\ \text{g solution}}{3.75\ \text{g AgNO}_3}$

$\dfrac{96.25\ \text{g H}_2\text{O}}{100.00\ \text{g solution}}$ and $\dfrac{100.00\ \text{g solution}}{96.25\ \text{g H}_2\text{O}}$

$\dfrac{3.75\ \text{g AgNO}_3}{96.25\ \text{g H}_2\text{O}}$ and $\dfrac{96.25\ \text{g H}_2\text{O}}{3.75\ \text{g AgNO}_3}$

(d) 4.25% Li_2SO_4 $\dfrac{4.25 \text{ g } Li_2SO_4}{100.00 \text{ g solution}}$ and $\dfrac{100.00 \text{ g solution}}{4.25 \text{ g } Li_2SO_4}$

$\dfrac{95.75 \text{ g } H_2O}{100.00 \text{ g solution}}$ and $\dfrac{100.00 \text{ g solution}}{95.75 \text{ g } H_2O}$

$\dfrac{4.25 \text{ g } Li_2SO_4}{95.75 \text{ g } H_2O}$ and $\dfrac{95.75 \text{ g } H_2O}{4.25 \text{ g } Li_2SO_4}$

47. (a) 5.36 ~~g glucose~~ $\times \dfrac{100.0 \text{ g solution}}{10.0 \text{ ~~g glucose~~}} = 53.6$ g solution

(b) 25.0 ~~g sucrose~~ $\times \dfrac{100.0 \text{ g solution}}{12.5 \text{ ~~g sucrose~~}} = 200$ g solution $(2.00 \times 10^2 \text{ g})$

49. (a) 85.0 ~~g solution~~ $\times \dfrac{2.00 \text{ g } FeBr_2}{100.0 \text{ ~~g solution~~}} = 1.70$ g $FeBr_2$

(b) 105.0 ~~g solution~~ $\times \dfrac{5.00 \text{ g } Na_2CO_3}{100.0 \text{ ~~g solution~~}} = 5.25$ g Na_2CO_3

51. (a) 250.0 ~~g solution~~ $\times \dfrac{99.10 \text{ g } H_2O}{100.0 \text{ ~~g solution~~}} = 247.8$ g H_2O

(b) 100.0 ~~g solution~~ $\times \dfrac{95.00 \text{ g } H_2O}{100.0 \text{ ~~g solution~~}} = 95.00$ g H_2O

Section 14.9 *Molar Concentration*

53. (a) MM of NaCl = 22.99 g + 35.45 g = 58.44 g/mol

100.0 mL solution = 0.1000 L solution

$\dfrac{1.50 \text{ ~~g NaCl~~}}{0.1000 \text{ L solution}} \times \dfrac{1 \text{ mol NaCl}}{58.45 \text{ ~~g NaCl~~}} = 0.257 \ M$ NaCl

(b) MM of $K_2Cr_2O_7$ = 2(39.10 g) + 2(52.00 g) + 7(16.00 g) = 294.20 g/mol

100.0 mL solution = 0.1000 L solution

$\dfrac{1.50 \text{ ~~g } K_2Cr_2O_7}{0.1000 \text{ L solution}} \times \dfrac{1 \text{ mol } K_2Cr_2O_7}{294.20 \text{ ~~g } K_2Cr_2O_7} = 0.0510 \ M \ K_2Cr_2O_7$

(c) MM of $CaCl_2$ = 40.08 g + 2(35.45 g) = 110.98 g/mol

125 mL solution = 0.125 L solution

$$\frac{5.55 \text{ g } CaCl_2}{0.125 \text{ L solution}} \times \frac{1 \text{ mol } CaCl_2}{110.98 \text{ g } CaCl_2} = 0.400 \text{ } M \text{ } CaCl_2$$

(d) MM of Na_2SO_4 = 2(22.99 g) + 32.07 g + 4(16.00 g) = 142.05 g/mol

125 mL solution = 0.125 L solution

$$\frac{5.55 \text{ g } Na_2SO_4}{0.125 \text{ L solution}} \times \frac{1 \text{ mol } Na_2SO_4}{142.05 \text{ g } Na_2SO_4} = 0.313 \text{ } M \text{ } Na_2SO_4$$

55.

Solution	Unit Factors	
(a) 0.100 M LiI	$\dfrac{0.100 \text{ mol LiI}}{1 \text{ L solution}}$ and	$\dfrac{1 \text{ L solution}}{0.100 \text{ mol LiI}}$
	$\dfrac{0.100 \text{ mol LiI}}{1000 \text{ mL solution}}$ and	$\dfrac{1000 \text{ mL solution}}{0.100 \text{ mol LiI}}$
(b) 0.100 M $NaNO_3$	$\dfrac{0.100 \text{ mol } NaNO_3}{1 \text{ L solution}}$ and	$\dfrac{1 \text{ L solution}}{0.100 \text{ mol } NaNO_3}$
	$\dfrac{0.100 \text{ mol } NaNO_3}{1000 \text{ mL solution}}$ and	$\dfrac{1000 \text{ mL solution}}{0.100 \text{ mol } NaNO_3}$
(c) 0.500 M K_2CrO_4	$\dfrac{0.500 \text{ mol } K_2CrO_4}{1 \text{ L solution}}$ and	$\dfrac{1 \text{ L solution}}{0.500 \text{ mol } K_2CrO_4}$
	$\dfrac{0.500 \text{ mol } K_2CrO_4}{1000 \text{ mL solution}}$ and	$\dfrac{1000 \text{ mL solution}}{0.500 \text{ mol } K_2CrO_4}$
(d) 0.500 M $ZnSO_4$	$\dfrac{0.500 \text{ mol } ZnSO_4}{1 \text{ L solution}}$ and	$\dfrac{1 \text{ L solution}}{0.500 \text{ mol } ZnSO_4}$
	$\dfrac{0.500 \text{ mol } ZnSO_4}{1000 \text{ mL solution}}$ and	$\dfrac{1000 \text{ mL solution}}{0.500 \text{ mol } ZnSO_4}$

57. (a) MM of NaF = 22.99 g + 19.00 g = 41.99 g/mol

$$10.0 \text{ g NaF} \times \frac{1 \text{ mol NaF}}{41.99 \text{ g NaF}} \times \frac{1 \text{ L solution}}{0.275 \text{ mol NaF}} = 0.866 \text{ L solution}$$

(b) MM of $CdCl_2$ = 112.41 g + 2(35.45 g) = 183.31 g/mol

$$10.0 \text{ g CdCl}_2 \times \frac{1 \text{ mol CdCl}_2}{183.31 \text{ g CdCl}_2} \times \frac{1 \text{ L solution}}{0.275 \text{ mol CdCl}_2} = 0.198 \text{ L solution}$$

(c) MM of K_2CO_3 = 2(39.10 g) + 12.01 g + 3(16.00 g) = 138.21 g/mol

$$10.0 \text{ g K}_2\text{CO}_3 \times \frac{1 \text{ mol K}_2\text{CO}_3}{138.21 \text{ g K}_2\text{CO}_3} \times \frac{1 \text{ L solution}}{0.408 \text{ mol K}_2\text{CO}_3} = 0.177 \text{ L solution}$$

(d) MM of $Fe(ClO_3)_3$ = 55.85 g + 3(35.45 g) + 9(16.00 g) = 306.20 g/mol

$$10.0 \text{ g Fe(ClO}_3)_3 \times \frac{1 \text{ mol Fe(ClO}_3)_3}{306.20 \text{ g Fe(ClO}_3)_3} \times \frac{1 \text{ L solution}}{0.408 \text{ mol Fe(ClO}_3)_3}$$

$$= 0.0800 \text{ L solution}$$

59. (a) MM of NaOH = 22.99 g + 16.00 g + 1.01 g = 40.00 g/mol

$$1.00 \text{ L solution} \times \frac{0.100 \text{ mol NaOH}}{1 \text{ L solution}} \times \frac{40.00 \text{ g NaOH}}{1 \text{ mol NaOH}} = 4.00 \text{ g NaOH}$$

(b) MM of $LiHCO_3$ = 6.94 g + 1.01 g + 12.01 g + 3(16.00 g) = 67.96 g/mol

$$1.00 \text{ L solution} \times \frac{0.100 \text{ mol LiHCO}_3}{1 \text{ L solution}} \times \frac{67.96 \text{ g LiHCO}_3}{1 \text{ mol LiHCO}_3} = 6.80 \text{ g LiHCO}_3$$

(c) MM of $CuCl_2$ = 63.55 g + 2(35.45 g) = 134.45 g/mol

$$25.0 \text{ mL solution} \times \frac{0.500 \text{ mol CuCl}_2}{1000 \text{ mL solution}} \times \frac{134.45 \text{ g CuCl}_2}{1 \text{ mol CuCl}_2} = 1.68 \text{ g CuCl}_2$$

(d) MM of $KMnO_4$ = 39.10 g + 54.94 g + 4(16.00 g) = 158.04 g/mol

$$25.0 \text{ mL solution} \times \frac{0.500 \text{ mol KMnO}_4}{1000 \text{ mL solution}} \times \frac{158.04 \text{ g KMnO}_4}{1 \text{ mol KMnO}_4} = 1.98 \text{ g KMnO}_4$$

61. MM of $CaSO_4$ = 40.08 g + 32.07 g + 4(16.00 g) = 136.15 g/mol

100 mL solution = 0.100 L solution

$$\frac{0.209 \text{ g } CaSO_4}{0.100 \text{ L solution}} \times \frac{1 \text{ mol } CaSO_4}{136.15 \text{ g } CaSO_4} = 0.0154 \text{ } M \text{ } CaSO_4$$

63. (a) $\dfrac{0.524 \text{ g glucose}}{10.483 \text{ g solution}} \times 100 = 5.00\%$

(b) MM of $C_6H_{12}O_6$ = 6(12.01 g) + 12(1.01 g) + 6(16.00 g) = 180.18 g/mol

10.0 mL solution = 0.0100 L solution

$$\frac{0.524 \text{ g } C_6H_{12}O_6}{0.0100 \text{ L solution}} \times \frac{1 \text{ mol } C_6H_{12}O_6}{180.18 \text{ g } C_6H_{12}O_6} = 0.291 \text{ } M \text{ } C_6H_{12}O_6$$

Section 14.10 *Molal Concentration*

65. (a) MM of KF = 39.10 g + 19.00 g = 58.10 g/mol

$$\frac{10.0 \text{ g KF}}{2.50 \text{ kg } H_2O} \times \frac{1 \text{ mol KF}}{58.10 \text{ g KF}} = 0.0688 \text{ } m \text{ KF}$$

(b) MM of $ZnSO_4$ = 65.39 g + 32.07 g + 4(16.00 g) = 161.46 g/mol

375 g H_2O = 0.375 kg H_2O

$$\frac{10.0 \text{ g } ZnSO_4}{0.375 \text{ kg } H_2O} \times \frac{1 \text{ mol } ZnSO_4}{161.46 \text{ g } ZnSO_4} = 0.165 \text{ } m \text{ } ZnSO_4$$

67. MM of $C_{12}H_{22}O_{11}$ = 12(12.01 g) + 22(1.01 g) + 11(16.00 g) = 342.34 g/mol

$$6.50 \text{ kg } H_2O \times \frac{2.00 \text{ mol } C_{12}H_{22}O_{11}}{1 \text{ kg } H_2O} \times \frac{342.34 \text{ g } C_{12}H_{22}O_{11}}{1 \text{ mol } C_{12}H_{22}O_{11}} = 4450 \text{ g } C_{12}H_{22}O_{11}$$

Section 14.11 *Colligative Properties*

69. MM of CH_3OH = 12.01 g + 4(1.01 g) + 16.00 g = 32.05 g/mol

500.0 g H_2O = 0.5000 kg H_2O

$$m = \frac{100.0 \text{ g } CH_3OH}{0.5000 \text{ kg } H_2O} \times \frac{1 \text{ mol } CH_3OH}{32.05 \text{ g } CH_3OH} = 6.24 \, m$$

$m \, K_f = \Delta T_f$ $\qquad\qquad \Delta T_f = 6.24 \, m \times \dfrac{1.86°C}{m} = 11.6°C$

Freezing point of solution: 0.0°C – 11.6°C = –11.6°C

71. $m \, K_f = \Delta T_f$ $\qquad\qquad \Delta T_f = [0.00°C - (-3.72°C)] = 3.72°C$

$m = \Delta T_f \times \dfrac{1}{K_f}$ $\qquad\qquad m = 3.72°C \times \dfrac{m}{1.86°C} = 2.00 \, m$

100.0 g H_2O = 0.1000 kg H_2O

$$\text{MM of unknown} = \frac{36.0 \text{ g}}{0.1000 \text{ kg } H_2O} \times \frac{1 \text{ kg } H_2O}{2.00 \text{ mol}} = 1.80 \times 10^2 \text{ g/mol}$$

73. $m \, K_f = \Delta T_f$ $\qquad\qquad \Delta T_f = [-117.3°C - (-119.8°C)] = 2.5°C$

$m = \Delta T_f \times \dfrac{1}{K_f}$ $\qquad\qquad m = 2.5°C \times \dfrac{m}{1.99°C} = 1.3 \, m$

100.0 g alcohol = 0.1000 kg alcohol

$$\text{MM of unknown} = \frac{4.50 \text{ g}}{0.1000 \text{ kg alcohol}} \times \frac{1 \text{ kg alcohol}}{1.3 \text{ mol}} = 35 \text{ g/mol}$$

General Exercises

75. The water droplets in the Tyndall effect serve as a colloidal dispersion. Thus, the water droplets must range in size from 1 to 100 nanometers.

77. (a) dispersed particles separate in a centrifuge $\qquad$ *colloid*
     (b) dispersed particles demonstrate the Tyndall effect $\qquad$ *colloid*
     (c) dispersed particles pass through a membrane $\qquad$ *solution*

79. $\dfrac{0.0019 \text{ g N}_2}{100 \text{ g blood}} \times \dfrac{4.68 \text{ atm}}{1.00 \text{ atm}} = 0.0089 \text{ g N}_2/100 \text{ g blood}$

81. 100 mL solution = 0.100 L solution

$$1.00 \text{ L solution} \times \dfrac{22.8 \text{ g SO}_2}{0.100 \text{ L solution}} = 228 \text{ g SO}_2$$

83. Air is less soluble in hot water than in cold tap water. Therefore, air leaves the solution and forms bubbles on the inside surface of the pan.

85. The polar –OH on an alcohol can hydrogen bond with water causing it to be soluble. As the nonpolar C_xH_y– portion of the molecule increases in size, the molecule overall becomes less polar and is eventually immiscible with water.

87. (a) In a 40% solution, ethyl alcohol is the solute and water is the solvent.
    (b) In a 95% solution, water is the solute and ethyl alcohol is the solvent.

89. MM of NaClO = 22.99 g + 35.45 g + 16.00 g = 74.44 g/mol

Solute: $5.25 \text{ g NaClO} \times \dfrac{1 \text{ mol NaClO}}{74.44 \text{ g NaClO}} = 0.0705 \text{ mol NaClO}$

Solution: $100 \text{ g bleach} \times \dfrac{1 \text{ mL bleach}}{1.04 \text{ g bleach}} \times \dfrac{1 \text{ L bleach}}{1000 \text{ mL bleach}} = 0.0962 \text{ L bleach}$

Molarity: $\dfrac{0.0705 \text{ mol NaClO}}{0.0962 \text{ L bleach}} = 0.733 \, M \text{ NaClO}$

91. Sugar dissolves in water to give single molecules, NaCl dissociates to give two ions, and $BaCl_2$ dissociates to give three ions. The freezing point depression is related to the number of particles dissolved in solution. Thus, NaCl (Na^+ and Cl^-) solute lowers the freezing point two times that of sugar, and $BaCl_2$ (Ba^{2+} and 2 Cl^-) lowers the freezing point three times that of sugar.

Acids and Bases

Section 15.1 *Properties of Acids and Bases*

1. <u>General Properties of Acids</u>
   (1) Acids have a sour taste.
   (2) Acids have a pH < 7.
   (3) Acids turn blue litmus paper red.

3.

<u>Food</u>	<u>pH</u>	<u>Classification</u>
(a) egg white	7.9	basic
(b) sour milk	6.2	acidic
(c) maple syrup	7.0	neutral
(d) lime juice	1.8	acidic
(e) champagne	3.8	acidic
(f) tomato juice	4.1	acidic

Section 15.2 *Arrhenius Acids and Bases*

5.

<u>Acid</u>	<u>Ionization</u>	<u>Strength</u>
(a) $HClO_3(aq)$	~ 100%	strong
(b) $HIO(aq)$	~ 1%	weak
(c) $HBr(aq)$	~ 100%	strong
(d) $HC_7H_5O_2(aq)$	~ 1%	weak

7.

<u>Formula</u>	<u>Classification</u>		<u>Formula</u>	<u>Classification</u>
(a) $HClO(aq)$	Arrhenius acid	(b)	$KOH(aq)$	Arrhenius base
(c) $K_2SO_4(aq)$	salt	(d)	$Sr(OH)_2(aq)$	Arrhenius base

9.

	<u>Arrhenius Acid</u>	<u>Arrhenius Base</u>
(a)	$HI(aq)$	$NaOH(aq)$
(b)	$HC_2H_3O_2(aq)$	$LiOH(aq)$

11.

	Salt	Acid	Base
(a)	$NaF(aq)$	HF	$NaOH$
(b)	$MgI_2(aq)$	HI	$Mg(OH)_2$
(c)	$Ca(NO_3)_2(aq)$	HNO_3	$Ca(OH)_2$
(d)	$Li_2CO_3(aq)$	H_2CO_3	$LiOH$

Neutralization Reactions:

(a) $HF(aq) + NaOH(aq) \rightarrow NaF(aq) + H_2O(l)$
(b) $2\,HI(aq) + Mg(OH)_2(s) \rightarrow MgI_2(aq) + 2\,H_2O(l)$
(c) $2\,HNO_3(aq) + Ca(OH)_2(aq) \rightarrow Ca(NO_3)_2(aq) + 2\,H_2O(l)$
(d) $H_2CO_3(aq) + 2\,LiOH(aq) \rightarrow Li_2CO_3(aq) + 2\,H_2O(l)$

13. (a) $2\,HNO_3(aq) + Ca(OH)_2(aq) \rightarrow Ca(NO_3)_2(aq) + 2\,H_2O(l)$
    (b) $H_2CO_3(aq) + Ba(OH)_2(aq) \rightarrow BaCO_3(aq) + 2\,H_2O(l)$

Section 15.3 *Brønsted–Lowry Acids and Bases*

15.

	Acid	Base			Acid	Base
(a)	$HC_2H_3O_2(aq)$	$LiOH(aq)$		(b)	$HBr(aq)$	$NaCN(aq)$

17.

	Acid	Base			Acid	Base
(a)	$HI(aq)$	$H_2O(l)$		(b)	$HC_2H_3O_2(aq)$	$HS^-(aq)$

19. (a) $HF(aq) + NaHS(aq) \rightarrow H_2S(aq) + NaF(aq)$
    (b) $HNO_2(aq) + NaC_2H_3O_2(aq) \rightarrow NaNO_2(aq) + HC_2H_3O_2(aq)$

Section 15.4 *Acid–Base Indicators*

21.

	pH	Methyl Red Color			pH	Methyl Red Color
(a)	3	red		(b)	7	yellow

23.

	pH	Phenolphthalein Color			pH	Phenolphthalein Color
(a)	7	colorless		(b)	11	pink

25. At pH 5, a portion of the methyl red indicator is yellow and a portion is red. Therefore, the color of the indicator is orange (yellow + red).

Section 15.5 *Acid–Base Titrations*

27. $HCl(aq) + NaOH(aq) \rightarrow NaCl(aq) + H_2O(l)$

$$22.15 \; \cancel{mL \; solution} \; \times \; \frac{0.155 \; mol \; NaOH}{1000 \; \cancel{mL \; solution}} \; = \; 0.00343 \; mol \; NaOH$$

$$0.00343 \; \cancel{mol \; NaOH} \; \times \; \frac{1 \; mol \; HCl}{1 \; \cancel{mol \; NaOH}} \; = \; 0.00343 \; mol \; HCl$$

$$\frac{0.00343 \; mol \; HCl}{25.0 \; \cancel{mL \; solution}} \; \times \; \frac{1000 \; \cancel{mL \; solution}}{1 \; L \; solution} \; = \; \frac{0.137 \; mol \; HCl}{1 \; L \; solution}$$

Molarity of hydrochloric acid: $0.137 \; M$ HCl

29. $H_3PO_4(aq) + 3 \; NaOH(aq) \rightarrow Na_3PO_4(aq) + 3 \; H_2O(l)$

$$34.45 \; \cancel{mL \; solution} \; \times \; \frac{0.210 \; mol \; NaOH}{1000 \; \cancel{mL \; solution}} \; = \; 0.00723 \; mol \; NaOH$$

$$0.00723 \; \cancel{mol \; NaOH} \; \times \; \frac{1 \; mol \; H_3PO_4}{3 \; \cancel{mol \; NaOH}} \; = \; 0.00241 \; mol \; H_3PO_4$$

$$\frac{0.00241 \; mol \; H_3PO_4}{50.0 \; \cancel{mL \; solution}} \; \times \; \frac{1000 \; \cancel{mL \; solution}}{1 \; L \; solution} \; = \; \frac{0.0482 \; mol \; H_3PO_4}{1 \; L \; solution}$$

Molarity of phosphoric acid: $0.0482 \; M$ H_3PO_4

31. $H_2SO_4(aq) + 2 \; KOH(aq) \rightarrow K_2SO_4(aq) + 2 \; H_2O(l)$

$$41.05 \; \cancel{mL \; solution} \; \times \; \frac{0.165 \; mol \; KOH}{1000 \; \cancel{mL \; solution}} \; = \; 0.00677 \; mol \; KOH$$

$$0.00677 \; \cancel{mol \; KOH} \; \times \; \frac{1 \; mol \; H_2SO_4}{2 \; \cancel{mol \; KOH}} \; = \; 0.00339 \; mol \; H_2SO_4$$

$$0.00339 \; \cancel{mol \; H_2SO_4} \; \times \; \frac{1000 \; mL \; solution}{0.122 \; \cancel{mol \; H_2SO_4}} \; = \; 27.8 \; mL \; H_2SO_4$$

33. (a) MM of HCl = 36.46 g/mol

$$\frac{6.00 \text{ mol HCl}}{1000 \text{ mL solution}} \times \frac{36.46 \text{ g HCl}}{1 \text{ mol HCl}} \times \frac{1 \text{ mL solution}}{1.10 \text{ g solution}} \times 100$$

$$= 19.9\% \text{ HCl}$$

(b) MM of $HC_2H_3O_2$ = 60.06 g/mol

$$\frac{1.00 \text{ mol } HC_2H_3O_2}{1000 \text{ mL solution}} \times \frac{60.06 \text{ g } HC_2H_3O_2}{1 \text{ mol } HC_2H_3O_2} \times \frac{1 \text{ mL solution}}{1.01 \text{ g solution}} \times 100$$

$$= 5.95\% \text{ } HC_2H_3O_2$$

(c) MM of HNO_3 = 63.02 g/mol

$$\frac{0.500 \text{ mol } HNO_3}{1000 \text{ mL solution}} \times \frac{63.02 \text{ g } HNO_3}{1 \text{ mol } HNO_3} \times \frac{1 \text{ mL solution}}{1.01 \text{ g solution}} \times 100$$

$$= 3.12\% \text{ } HNO_3$$

(d) MM of H_2SO_4 = 98.09 g/mol

$$\frac{3.00 \text{ mol } H_2SO_4}{1000 \text{ mL solution}} \times \frac{98.09 \text{ g } H_2SO_4}{1 \text{ mol } H_2SO_4} \times \frac{1 \text{ mL solution}}{1.18 \text{ g solution}} \times 100$$

$$= 24.9\% \text{ } H_2SO_4$$

Section 15.6 *Acid–Base Standardization*

35. $2 HNO_3(aq) + Na_2CO_3(s) \rightarrow 2 NaNO_3(aq) + H_2O(l) + CO_2(g)$

MM of Na_2CO_3 = 105.99 g/mol

$$0.689 \text{ g } Na_2CO_3 \times \frac{1 \text{ mol } Na_2CO_3}{105.99 \text{ g } Na_2CO_3} \times \frac{2 \text{ mol } HNO_3}{1 \text{ mol } Na_2CO_3} = 0.0130 \text{ mol } HNO_3$$

$$\frac{0.0130 \text{ mol } HNO_3}{41.25 \text{ mL solution}} \times \frac{1000 \text{ mL solution}}{1 \text{ L solution}} = \frac{0.315 \text{ mol } HNO_3}{1 \text{ L solution}}$$

Molarity of nitric acid: $0.315 M \text{ } HNO_3$

37. $2\,HCl(aq) + Na_2C_2O_4(s) \rightarrow H_2C_2O_4(aq) + 2\,NaCl(aq)$

MM of $Na_2C_2O_4$ = 134.00 g/mol

$$1.550\;\cancel{g\,Na_2C_2O_4} \times \frac{1\;\cancel{mol\,Na_2C_2O_4}}{134.00\;\cancel{g\,Na_2C_2O_4}} \times \frac{2\;mol\,HCl}{1\;\cancel{mol\,Na_2C_2O_4}} = 0.02313\;mol\,HCl$$

$$\frac{0.02313\;mol\,HCl}{20.95\;\cancel{mL\,solution}} \times \frac{1000\;\cancel{mL\,solution}}{1\;L\,solution} = \frac{1.104\;mol\,HCl}{1\;L\,solution}$$

Molarity of hydrochloric acid: 1.104 M HCl

39. $H_2C_2O_4(s) + 2\,LiOH(aq) \rightarrow Li_2C_2O_4(aq) + 2\,H_2O(l)$

MM of $H_2C_2O_4$ = 90.04 g/mol

$$0.627\;\cancel{g\,H_2C_2O_4} \times \frac{1\;\cancel{mol\,H_2C_2O_4}}{90.04\;\cancel{g\,H_2C_2O_4}} \times \frac{2\;mol\,LiOH}{1\;\cancel{mol\,H_2C_2O_4}} = 0.0139\;mol\,LiOH$$

$$0.0139\;\cancel{mol\,LiOH} \times \frac{1000\;mL\,solution}{0.479\;\cancel{mol\,LiOH}} = 29.0\;mL\,solution$$

Volume of lithium hydroxide: 29.0 mL LiOH

41. $HAsc(s) + NaOH(aq) \rightarrow NaAsc(aq) + H_2O(l)$

$$30.95\;\cancel{mL\,solution} \times \frac{0.176\;\cancel{mol\,NaOH}}{1000\;\cancel{mL\,solution}} \times \frac{1\;mol\,HAsc}{1\;\cancel{mol\,NaOH}} = 0.00545\;mol\,HAsc$$

$$\text{MM of vitamin C} = \frac{0.959\;g\,HAsc}{0.00545\;mol\,HAsc} = 176\;g/mol$$

43. 1 mole of alanine = 1 mole NaOH

$$21.05\;\cancel{mL\,solution} \times \frac{0.145\;\cancel{mol\,NaOH}}{1000\;\cancel{mL\,solution}} \times \frac{1\;mol\,alanine}{1\;\cancel{mol\,NaOH}}$$

$$= 0.00305\;mol\,alanine$$

$$\text{MM of alanine} = \frac{0.272\;g\,alanine}{0.00305\;mol\,alanine} = 89.2\;g/mol$$

Section 15.7 *Ionization of Water*

45.   (a)   Simplified ionization equation: $H_2O(l) \rightarrow H^+(aq) + OH^-(aq)$
     (b)   Ionization constant expression: $K_w = [H^+][OH^-]$
     (c)   Ionization constant for water at 25°C: $K_w = 1.0 \times 10^{-14}$

47. $[OH^-] = \dfrac{1.0 \times 10^{-14}}{[H^+]}$

     (a) $[OH^-] = \dfrac{1.0 \times 10^{-14}}{0.025} = 4.0 \times 10^{-13}$

     (b) $[OH^-] = \dfrac{1.0 \times 10^{-14}}{1.7 \times 10^{-5}} = 5.9 \times 10^{-10}$

49. $[H^+] = \dfrac{1.0 \times 10^{-14}}{[OH^-]}$

     (a) $[H^+] = \dfrac{1.0 \times 10^{-14}}{0.0016} = 6.3 \times 10^{-12}$

     (b) $[H^+] = \dfrac{1.0 \times 10^{-14}}{0.000\,29} = 3.4 \times 10^{-11}$

Section 15.8 *The pH Concept*

51. $pH = -\log[H^+]$

     (a) $pH = -\log 0.001 = -\log 10^{-3}$
        $pH = -(-3) = 3$

     (b) $pH = -\log 0.000\,01 = -\log 10^{-5}$
        $pH = -(-5) = 5$

53. $[H^+] = 10^{-pH}$

     (a) $[H^+] = 10^{-6} = 0.000\,001\,M\ (1 \times 10^{-6}\,M)$

     (b) $[H^+] = 10^{-8} = 0.000\,000\,01\,M\ (1 \times 10^{-8}\,M)$

Section 15.9 *Advanced pH Calculations*

55.　(a)　$pH = -\log 0.000\,0079 = -\log 7.9 \times 10^{-6}$
　　　　$pH = -\log 7.9 - \log 10^{-6}$
　　　　$pH = -0.90 - (-6) = 5.10$

　　(b)　$pH = -\log 0.000\,000\,39 = -\log 3.9 \times 10^{-7}$
　　　　$pH = -\log 3.9 - \log 10^{-7}$
　　　　$pH = -0.59 - (-7) = 6.41$

57.　(a)　$[H^+] = 10^{-1.80} = 10^{0.20} \times 10^{-2}$
　　　　$[H^+] = 1.6 \times 10^{-2}\,M = 0.016\,M$

　　(b)　$[H^+] = 10^{-4.75} = 10^{0.25} \times 10^{-5}$
　　　　$[H^+] = 1.8 \times 10^{-5}\,M = 0.000\,018\,M$

59.　(a)　$[OH^-] = 0.11\,M$

$$[H^+] = \frac{1.0 \times 10^{-14}}{[OH^-]} = \frac{1.0 \times 10^{-14}}{0.11} = 9.1 \times 10^{-14}$$

　　　　$pH = -\log (9.1 \times 10^{-14})$
　　　　$pH = -\log 9.1 - \log 10^{-14}$
　　　　$pH = -0.96 - (-14) = 13.04$

　　(b)　$[OH^-] = 0.000\,55\,M = 5.5 \times 10^{-4}\,M$

$$[H^+] = \frac{1.0 \times 10^{-14}}{[OH^-]} = \frac{1.0 \times 10^{-14}}{5.5 \times 10^{-4}} = 1.8 \times 10^{-11}$$

　　　　$pH = -\log (1.8 \times 10^{-11})$
　　　　$pH = -\log 1.8 - \log 10^{-11}$
　　　　$pH = -0.26 - (-11) = 10.74$

61.　(a)　$pH = 0.90$
　　　　$[H^+] = 10^{-0.90} = 10^{0.10} \times 10^{-1}$
　　　　$[H^+] = 1.3 \times 10^{-1}\,M = 0.13\,M$

$$[OH^-] = \frac{1.0 \times 10^{-14}}{[H^+]} = \frac{1.0 \times 10^{-14}}{0.13} = 7.7 \times 10^{-14}$$

　　(b)　$pH = 1.62$
　　　　$[H^+] = 10^{-1.62} = 10^{0.38} \times 10^{-2}$
　　　　$[H^+] = 2.4 \times 10^{-2}\,M = 0.024\,M$

$$[OH^-] = \frac{1.0 \times 10^{-14}}{[H^+]} = \frac{1.0 \times 10^{-14}}{0.024} = 4.2 \times 10^{-13}$$

Section 15.10 *Strong and Weak Electrolytes*

63.
	Solution	Ionization
(a)	strong acids	highly ionized
(b)	strong bases	highly ionized
(c)	soluble ionic compounds	highly ionized

65.
	Solution	Electrolyte
(a)	$H_2CO_3(aq)$	weak
(b)	$H_2SO_4(aq)$	strong
(c)	$HI(aq)$	strong
(d)	$HNO_2(aq)$	weak

67.
	Solution	Electrolyte
(a)	$ZnCO_3(s)$	weak
(b)	$Sr(NO_3)_2(aq)$	strong
(c)	$K_2SO_4(aq)$	strong
(d)	$PbI_2(s)$	weak

69.
	Solution	Electrolyte	Aqueous Solution
(a)	$HF(aq)$	weak	$HF(aq)$
(b)	$HBr(aq)$	strong	$H^+(aq)$ and $Br^-(aq)$
(c)	$HNO_3(aq)$	strong	$H^+(aq)$ and $NO_3^-(aq)$
(d)	$HNO_2(aq)$	weak	$HNO_2(aq)$

71.
	Solution	Electrolyte	Aqueous Solution
(a)	$AgF(aq)$	strong	$Ag^+(aq)$ and $F^-(aq)$
(b)	$AgI(s)$	weak	$AgI(s)$
(c)	$Hg_2Cl_2(s)$	weak	$Hg_2Cl_2(s)$
(d)	$NiCl_2(aq)$	strong	$Ni^{2+}(aq)$ and $2\,Cl^-(aq)$

Section 15.11 *Net Ionic Equations*

73.
(1) Complete and balance the nonionized equation.
(2) Convert the nonionized equation to the total ionic equation.
(3) Cancel spectator ions to obtain the net ionic equation.
(4) Check ($\sqrt{}$) each ion or atom on both sides of the equation.

75.
(a) Nonionized equation:
$$HCl(aq) + KOH(aq) \rightarrow KCl(aq) + H_2O(l)$$

Total ionic equation:
$$H^+(aq) + Cl^-(aq) + K^+(aq) + OH^-(aq) \rightarrow K^+(aq) + Cl^-(aq) + H_2O(l)$$

Net ionic equation:
$$H^+(aq) + OH^-(aq) \rightarrow H_2O(l)$$

(b) Nonionized equation:
$$2\,HC_2H_3O_2(aq) + Ca(OH)_2(aq) \rightarrow Ca(C_2H_3O_2)_2(aq) + 2\,H_2O(l)$$

Total ionic equation:
$$2\,HC_2H_3O_2(aq) + Ca^{2+}(aq) + 2\,OH^-(aq) \rightarrow$$
$$Ca^{2+}(aq) + 2\,C_2H_3O_2^-(aq) + 2\,H_2O(l)$$

Net ionic equation:
$$HC_2H_3O_2(aq) + OH^-(aq) \rightarrow C_2H_3O_2^-(aq) + H_2O(l)$$

77. (a) Nonionized equation:
$$AgNO_3(aq) + KI(aq) \rightarrow AgI(s) + KNO_3(aq)$$

Total ionic equation:
$$Ag^+(aq) + NO_3^-(aq) + K^+(aq) + I^-(aq) \rightarrow K^+(aq) + NO_3^-(aq) + AgI(s)$$

Net ionic equation:
$$Ag^+(aq) + I^-(aq) \rightarrow AgI(s)$$

(b) Nonionized equation:
$$BaCl_2(aq) + K_2CrO_4(aq) \rightarrow BaCrO_4(s) + 2\,KCl(aq)$$

Total ionic equation:
$$Ba^{2+}(aq) + 2\,Cl^-(aq) + 2\,K^+(aq) + CrO_4^{2-}(aq) \rightarrow$$
$$BaCrO_4(s) + 2\,K^+(aq) + 2\,Cl^-(aq)$$

Net ionic equation:
$$Ba^{2+}(aq) + CrO_4^{2-}(aq) \rightarrow BaCrO_4(s)$$

General Exercises

79. The color of methyl red is yellow at pH 7, while phenolphthalein is colorless. Therefore, the color of the water with both indicators is *yellow*.

81. The color of methyl orange is red at pH 3.2 and yellow at pH 4.4. Therefore, the color of a solution at pH 3.8 is *orange* (red + yellow).

83. Notice that NaH_2PO_4 is a *proton acceptor* in the first reaction and a *proton donor* in the second reaction. Thus, NaH_2PO_4 is amphiprotic.

85. $42.10 \text{ mL solution} \times \dfrac{0.100 \text{ mol Ba(OH)}_2}{1000 \text{ mL solution}} \times \dfrac{1 \text{ mol cream of tartar}}{1 \text{ mol Ba(OH)}_2}$

$$= 0.00421 \text{ mol Ba(OH)}_2$$

$$\text{MM cream of tartar} = \dfrac{0.791 \text{ g cream of tartar}}{0.00421 \text{ mol cream of tartar}} = 188 \text{ g/mol}$$

87. $\text{HCl(aq)} \rightarrow \text{H}^+\text{(aq)} + \text{Cl}^-\text{(aq)}$

$0.50 \, M \text{ HCl} \rightarrow 0.50 \, M \text{ H}^+$

$\text{pH} = -\log[\text{H}^+]$
$\text{pH} = -\log 0.50$
$\text{pH} = 0.30$

89. Nonionized equation:
$\text{NH}_4\text{OH(aq)} + \text{HC}_2\text{H}_3\text{O}_2\text{(aq)} \rightarrow \text{NH}_4\text{C}_2\text{H}_3\text{O}_2\text{(aq)} + \text{H}_2\text{O(l)}$

Total ionic equation:
$\text{NH}_4\text{OH(aq)} + \text{HC}_2\text{H}_3\text{O}_2\text{(aq)} \rightarrow \text{NH}_4^+\text{(aq)} + \text{C}_2\text{H}_3\text{O}_2^-\text{(aq)} + \text{H}_2\text{O(l)}$

Net ionic equation:
$\text{NH}_4\text{OH(aq)} + \text{HC}_2\text{H}_3\text{O}_2\text{(aq)} \rightarrow \text{NH}_4^+\text{(aq)} + \text{C}_2\text{H}_3\text{O}_2^-\text{(aq)} + \text{H}_2\text{O(l)}$

Chemical Equilibrium

Section 16.1 *Collision Theory*

1.   Reaction Rate Factors
      (1) frequency of molecular collisions
      (2) energy of molecular collisions
      (3) orientation of colliding molecules (collision geometry)

3.   Effective Collision Geometry

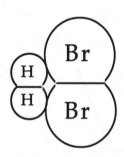

5.   Reaction Change   Effect on the Rate of Reaction
      (a) Increase the concentration   The rate of reaction *increases*
           of a reactant   because more collisions occur.

      (b) Decrease the temperature   The rate of reaction *decreases*
           of the reaction   because both collision frequency
                                         and collision energy decrease.

      (c) Add a catalyst   The rate of reaction *increases*
                                         because collision geometry is
                                         more effective.

7.   In a coal mine, fine particles of coal dust in the air provide a greater surface
      area for reaction. In the barbecue, large briquettes have less surface area to
      contact the oxygen in air.

9. $PCl_5(g) + heat \rightarrow PCl_3(g) + Cl_2(g)$

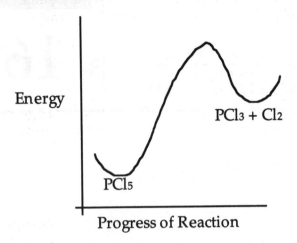

11. $H_2(g) + I_2(g) + heat \rightarrow 2\,HI(g)$

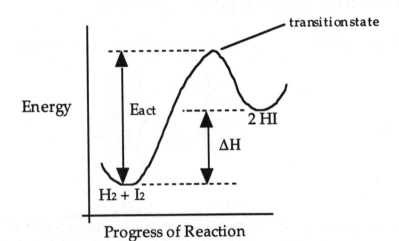

13.   A catalyst is a substance that allows a reaction to proceed faster by lowering the energy of activation (E_{act}).

15.   Since the reaction is endothermic, more energy is required for the reaction to proceed in the forward direction toward products. Thus, the E_{act} is greater in the forward direction than in the reverse direction.

Section 16.3 *The Chemical Equilibrium Concept*

17. (a) The rate of the forward reaction is measured by the rate of change in the decrease in concentration of reactant(s) in a given unit of time.

$$\text{rate of a forward reaction } = \frac{\text{decrease [reactant]}}{\text{unit time}}$$

(b) The rate of the forward reaction is measured by the rate of change in the increase in concentration of product(s) in a given unit of time.

$$\text{rate of a forward reaction } = \frac{\text{increase [product]}}{\text{unit time}}$$

19. Both the statements (a) and (b) are true regarding the general equilibrium expression. That is, the equilibrium expression for a given reaction can be determined experimentally and derived theoretically.

Section 16.4 *General Equilibrium Constant,* K_{eq}

21. **Equilibrium Reaction**   **Equilibrium Constant Expression**

(a) $2\,A \rightleftarrows C$ $\qquad\qquad\qquad\qquad K_{eq} = \dfrac{[C]}{[A]^2}$

(b) $A + 2\,B \rightleftarrows 3\,C$ $\qquad\qquad\quad K_{eq} = \dfrac{[C]^3}{[A]\,[B]^2}$

(c) $2\,A + 3\,B \rightleftarrows 4\,C + D$ $\qquad\quad K_{eq} = \dfrac{[C]^4\,[D]}{[A]^2\,[B]^3}$

23. Since the amount of any solid present has no effect on the equilibrium of a gaseous state reaction, any substance in the solid state will not appear in the equilibrium expression for a gaseous state reaction.

25. **Equilibrium Reaction**   **Equilibrium Constant Expression**

(a) $H_2(g) + F_2(g) \rightleftarrows 2\,HF(g)$ $\qquad\qquad K_{eq} = \dfrac{[HF]^2}{[H_2]\,[F_2]}$

(b) $4\,NH_3(g) + 7\,O_2(g) \rightleftarrows 4\,NO_2(g) + 6\,H_2O(g)$ $\;\;K_{eq} = \dfrac{[NO_2]^4\,[H_2O]^6}{[NH_3]^4\,[O_2]^7}$

(c) $ZnCO_3(s) \rightleftarrows ZnO(s) + CO_2(g)$ $\qquad\quad K_{eq} = [CO_2]$

27. $2\,SO_2(g) + O_2(g) \rightleftarrows 2\,SO_3(g)$

$$[SO_2] = 1.75\,M \qquad [O_2] = 1.50\,M \qquad [SO_3] = 2.25\,M$$

$$K_{eq} = \frac{[SO_3]^2}{[SO_2]^2\,[O_2]} = \frac{(2.25)^2}{(1.75)^2\,(1.50)} = 1.10$$

Section 16.5 *Gaseous State Equilibria Shifts*

29. $\qquad N_2O_4(g) \;+\; heat \;\;\rightleftarrows\;\; 2\,NO_2(g)$

(a) On hot days, the increase in heat shifts the equilibrium to the right producing more NO_2. Thus, the NO_2 concentration *increases*.

(b) On cool days, the decrease in heat shifts the equilibrium to the left producing less NO_2. Thus, the NO_2 concentration *decreases*.

31. $\qquad C(s) \;+\; CO_2(g) \;+\; heat \;\;\rightleftarrows\;\; 2\,CO(g)$

(a)	[CO$_2$] decreases	shift left	(b)	[CO] decreases	shift right
(c)	charcoal is added	no shift	(d)	CO$_2$(g) is added	shift right
(e)	temp. increases	shift right	(f)	temp. decreases	shift left
(g)	pressure increases	shift left	(h)	pressure decreases	shift right

33. $\qquad SO_2(g) \;+\; NO_2(g) \;\;\rightleftarrows\;\; SO_3(g) \;+\; NO(g) \;+\; heat$

(a)	[SO$_2$] decreases	shift left	(b)	[SO$_3$] decreases	shift right
(c)	[NO$_2$] increases	shift right	(d)	[NO] increases	shift left
(e)	temp. decreases	shift right	(f)	pressure increases	no shift
(g)	pressure decreases	no shift	(h)	ultraviolet light	no shift

Section 16.6 *Ionization Equilibrium Constant,* K_i

35. $\qquad$ <u>Equilibrium Reaction</u> $\qquad\qquad$ <u>Equilibrium Constant Expression</u>

(a) $HCHO_2(aq) \rightleftarrows H^+(aq) + CHO_2^-(aq) \qquad K_i = \dfrac{[H^+]\,[CHO_2^-]}{[HCHO_2]}$

(b) $H_2C_2O_4(aq) \rightleftarrows H^+(aq) + HC_2O_4^-(aq) \qquad K_i = \dfrac{[H^+]\,[HC_2O_4^-]}{[H_2C_2O_4]}$

(c) $H_3C_6H_5O_7(aq) \rightleftarrows H^+(aq) + H_2C_6H_5O_7^-(aq) \quad K_i = \dfrac{[H^+]\,[H_2C_6H_5O_7^-]}{[H_3C_6H_5O_7]}$

37. $HNO_2(aq) \rightleftharpoons H^+(aq) + NO_2^-(aq)$

$$[H^+] = [NO_2^-] = 7.5 \times 10^{-3} M \qquad [HNO_2] = 0.125 M$$

$$K_i = \frac{[H^+][NO_2^-]}{[HNO_2]} = \frac{(7.5 \times 10^{-3})(7.5 \times 10^{-3})}{(0.125)} = 4.5 \times 10^{-4}$$

39. $HF(aq) \rightleftharpoons H^+(aq) + F^-(aq)$

$$[H^+] = 10^{-pH} = 10^{-2.00} = 0.010 M$$

$$[H^+] = [F^-] = 0.010 M \qquad [HF] = 0.139 M$$

$$K_i = \frac{[H^+][F^-]}{[HF]} = \frac{(0.010)(0.010)}{(0.139)} = 7.2 \times 10^{-4}$$

Section 16.7 *Weak Acid–Base Equilibria Shifts*

41. $HF(aq) \rightleftharpoons H^+(aq) + F^-(aq)$

(a)	increase [HF]	shift right	(b)	increase [H$^+$]	shift left
(c)	decrease [HF]	shift left	(d)	decrease [F$^-$]	shift right
(e)	add NaF solid	shift left	(f)	add HCl gas	shift left
(g)	add NaOH solid	shift right	(h)	increase pH	shift right

43. $HC_2H_3O_2(aq) \rightleftharpoons H^+(aq) + C_2H_3O_2^-(aq)$

(a)	increase [HC$_2$H$_3$O$_2$]	shift right	(b)	increase [H$^+$]	shift left
(c)	decrease [HC$_2$H$_3$O$_2$]	shift left	(d)	decrease [C$_2$H$_3$O$_2^-$]	shift right
(e)	add NaC$_2$H$_3$O$_2$ solid	shift left	(f)	add NaCl solid	no shift
(g)	add NaOH solid	shift right	(h)	increase pH	shift right

Section 16.8 *Solubility Product Equilibrium Constant,* K$_{sp}$

45.

Reaction	Solubility Product Expression
(a) $AgI(s) \rightleftharpoons Ag^+(aq) + I^-(aq)$	$K_{sp} = [Ag^+][I^-]$
(b) $Ag_2CrO_4(s) \rightleftharpoons 2\,Ag^+(aq) + CrO_4^{2-}(aq)$	$K_{sp} = [Ag^+]^2[CrO_4^{2-}]$
(c) $Ag_3PO_4(s) \rightleftharpoons 3\,Ag^+(aq) + PO_4^{3-}(aq)$	$K_{sp} = [Ag^+]^3[PO_4^{3-}]$

47. $CoS(s) \rightleftarrows Co^{2+}(aq) + S^{2-}(aq)$

$$[Co^{2+}] = [S^{2-}] = 7.7 \times 10^{-11}$$

$$K_{sp} = [Co^{2+}][S^{2-}] = (7.7 \times 10^{-11})(7.7 \times 10^{-11}) = 5.9 \times 10^{-21}$$

49. $Zn_3(PO_4)_2(s) \rightleftarrows 3 Zn^{2+}(aq) + 2 PO_4^{3-}(aq)$

If $[Zn^{2+}] = 1.5 \times 10^{-7}$ and 2 PO_4^{3-} are produced for every 3 Zn^{2+},

then, $[PO_4^{3-}] = \frac{2}{3}[Zn^{2+}] = \frac{2}{3}(1.5 \times 10^{-7}) = 1.0 \times 10^{-7}$

$$K_{sp} = [Zn^{2+}]^3[PO_4^{3-}]^2 = (1.5 \times 10^{-7})^3(1.0 \times 10^{-7})^2 = 3.4 \times 10^{-35}$$

51. $CaCO_3(s) \rightleftarrows Ca^{2+}(aq) + CO_3^{2-}(aq)$

$$[Ca^{2+}][CO_3^{2-}] = K_{sp} = 3.8 \times 10^{-9}$$

Since $[Ca^{2+}] = [CO_3^{2-}]$,

then, $[Ca^{2+}]^2 = 3.8 \times 10^{-9}$

thus, $[Ca^{2+}] = 6.2 \times 10^{-5}$

$CaC_2O_4(s) \rightleftarrows Ca^{2+}(aq) + C_2O_4^{2-}(aq)$

$$[Ca^{2+}][C_2O_4^{2-}] = K_{sp} = 2.3 \times 10^{-9}$$

Since $[Ca^{2+}] = [C_2O_4^{2-}]$,

then, $[Ca^{2+}]^2 = 2.3 \times 10^{-9}$

thus, $[Ca^{2+}] = 4.8 \times 10^{-5}$

The calcium ion concentration in a saturated solution of $CaCO_3$ is slightly greater than in a saturated solution of CaC_2O_4.

Section 16.9 *Solubility Equilibria Shifts*

53. $\quad$ $Ca_3(PO_4)_2(s) \quad \rightleftarrows \quad 3\,Ca^{2+}(aq) \;+\; 2\,PO_4^{3-}(aq)$

(a)	increase $[Ca^{2+}]$	shift left	(b)	increase $[PO_4^{3-}]$	shift left
(c)	decrease $[Ca^{2+}]$	shift right	(d)	decrease $[PO_4^{3-}]$	shift right
(e)	add solid $Ca(NO_3)_2$	shift left	(f)	add solid KNO_3	no shift
(g)	add solid $Ca_3(PO_4)_2$	no shift	(h)	add H^+	shift right

55. $\quad$ $Cu(OH)_2(s) \quad \rightleftarrows \quad Cu^{2+}(aq) \;+\; 2\,OH^-(aq)$

(a)	increase $[Cu^{2+}]$	shift left	(b)	increase $[OH^-]$	shift left
(c)	decrease $[Cu^{2+}]$	shift right	(d)	decrease $[OH^-]$	shift right
(e)	add solid $Cu(OH)_2$	no shift	(f)	add solid $NaOH$	shift left
(g)	add solid $NaCl$	no shift	(h)	decrease pH	shift right

General Exercises

57. A swinging pendulum in a clock represents a *dynamic process* as it is in constant motion. The moving pendulum represents a *reversible process* as it swings back and forth in opposite directions.

59. With regard to rate of reaction, a system at equilibrium has the following characteristic: *the rate of the forward reaction is equal to the rate of the reverse reaction.*

61. $N_2O_4(g) \;\rightleftarrows\; 2\,NO_2(g)$

$$[N_2O_4] = 4.5 \times 10^{-5}\, M \qquad [NO_2] = 3.0 \times 10^{-3}\, M$$

$$K_{eq} \;=\; \frac{[NO_2]^2}{[N_2O_4]} \;=\; \frac{(3.0 \times 10^{-3})^2}{(4.5 \times 10^{-5})} \;=\; 0.20$$

63. $\quad$ $H_2O(l) \quad \rightleftarrows \quad H^+(aq) \;+\; OH^-(aq)$

(a)	increase $[H^+]$	shift left	(b)	decrease $[OH^-]$	shift right
(c)	increase pH	shift right	(d)	decrease pH	shift left

65. $Fe(OH)_3(s) \rightleftarrows Fe^{3+}(aq) + 3\,OH^-(aq)$

Aqueous HCl ionizes to give $H^+(aq)$ and $Cl^-(aq)$. The $H^+(aq)$ can react with $OH^-(aq)$ to form H_2O. This decreases the $OH^-(aq)$ concentration which shifts the equilibrium to the right, thus allowing more dissociation of $Fe(OH)_3(s)$.

In summary, $Fe(OH)_3$ is more soluble in aqueous HCl than in water owing to the following reaction:

$$Fe(OH)_3(s) + 3\,H^+(aq) \rightleftarrows Fe^{3+}(aq) + 3\,H_2O(l)$$

67. $Ca(OH)_2(s) \rightleftarrows Ca^{2+}(aq) + 2\,OH^-(aq)$

$pH = 12.35$
$[H^+] = 10^{-12.35} = 10^{0.65} \times 10^{-13}$
$[H^+] = 4.5 \times 10^{-13}\,M$

$$[OH^-] = \frac{1.0 \times 10^{-14}}{[H^+]} = \frac{1.0 \times 10^{-14}}{4.5 \times 10^{-13}} = 2.2 \times 10^{-2}$$

If $[OH^-] = 2.2 \times 10^{-2}$ and 1 Ca^{2+} ionizes for every 2 OH^-,

then, $[Ca^{2+}] = \frac{1}{2}[OH^-] = \frac{1}{2}(2.2 \times 10^{-2}) = 1.1 \times 10^{-2}$

Therefore,

$$K_{sp} = [Ca^{2+}][OH^-]^2 = (1.1 \times 10^{-2})(2.2 \times 10^{-2})^2 = 5.3 \times 10^{-6}$$

Oxidation and Reduction

Section 17.1 *Oxidation Numbers*

1.

	Metal	Ox No		Metal	Ox No
(a)	Mg	0	(b)	Mn	0
(c)	K	0	(d)	Zn	0

(*Note:* All elements in the free state have an oxidation number of zero.)

3.

	Cation	Ox No		Cation	Ox No
(a)	Sr^{2+}	+2	(b)	Sc^{3+}	+3
(c)	Ti^{4+}	+4	(d)	Ag^+	+1

5.
	Compound	Oxidation Number of Silicon

(a) SiO_2

ox no Si + 2(ox no O) = 0
ox no Si + 2(–2) = 0
ox no Si + (–4) = 0
ox no Si = +4

(b) Si_2H_6

2(ox no Si) + 6(ox no H) = 0
2(ox no Si) + 6(–1) = 0
2(ox no Si) + (–6) = 0
2(ox no Si) = +6
ox no Si = +3

(c) Si_3N_4

3(ox no Si) + 4(ox no N) = 0
3(ox no Si) + 4(–3) = 0
3(ox no Si) + (–12) = 0
3(ox no Si) = +12
ox no Si = +4

(d) $CaSiO_3$

ox no Ca + ox no Si + 3(ox no O) = 0
+2 + ox no Si + 3(–2) = 0
+2 + ox no Si + (–6) = 0
ox no Si = +4

7.

Ion	Oxidation Number of Carbon

(a) CO_3^{2-}

ox no C + 3(ox no O) = –2
ox no C + 3(–2) = –2
ox no C + (–6) = –2
ox no C = –2 +6
ox no C = +4

(b) HCO_3^-

ox no H + ox no C + 3(ox no O) = –1
+1 + ox no C + 3(–2) = –1
+1 + ox no C + (–6) = –1
ox no C –5 = –1
ox no C = –1 +5
ox no C = +4

(c) CN^-

ox no C + ox no N = –1
ox no C + (–3) = –1
ox no C = –1 +3
ox no C = +2

(d) $CNO-$

ox no C + ox no N + ox no O = –1
ox no C + (–3) + (–2) = –1
ox no C + (–5) = –1
ox no C = –1 +5
ox no C = +4

Section 17.2 *Oxidation–Reduction Reactions*

9. (a) A redox process characterized by electron loss is termed *oxidation*.
   (b) A redox process characterized by electron gain is termed *reduction*.

11. (a)
$$Mn_{(s)} + O_{2(g)} \rightarrow MnO_{2(s)}$$

Oxidized: Mn   Reduced: O_2

(b)
$$S_{(s)} + O_{2(g)} \rightarrow SO_{2(g)}$$

Oxidized: S   Reduced: O_2

13. (a)

$$CuO_{(s)} + H_{2(g)} \rightarrow Cu_{(s)} + H_2O_{(l)}$$

(with oxidation numbers: CuO as $+2$ and 0 above, Cu as 0, H_2O as $+1$)

Oxidizing agent: CuO Reducing agent: H_2

(b)

$$PbO_{(s)} + CO_{(g)} \rightarrow Pb_{(s)} + CO_{2(g)}$$

(with oxidation numbers: PbO as $+2$, CO as $+2$ and 0, CO_2 as $+4$)

Oxidizing agent: PbO Reducing agent: CO

15. (a)

$$Al_{(s)} + Cr^{3+}_{(aq)} \rightarrow Al^{3+}_{(aq)} + Cr_{(s)}$$

(with oxidation numbers: Al as 0, Cr^{3+} as $+3$, Al^{3+} as $+3$, Cr as 0)

Oxidized: Al Reduced: Cr^{3+}

(b)

$$F_{2(g)} + 2\,Cl^-_{(aq)} \rightarrow 2\,F^-_{(aq)} + Cl_{2(g)}$$

(with oxidation numbers: F_2 as 0, Cl^- as -1, F^- as -1, Cl_2 as 0)

Oxidized: Cl^- Reduced: F_2

17. (a)

$$Cr^{2+}_{(aq)} + AgI_{(s)} \rightarrow Cr^{3+}_{(aq)} + Ag_{(s)} + I^-_{(aq)}$$

(with oxidation numbers: Cr^{2+} as $+2$, AgI as $+1$, Cr^{3+} as $+3$, Ag as 0)

Oxidizing agent: AgI Reducing agent: Cr^{2+}

(b)

$$Sn^{2+}_{(aq)} + 2\,Hg^{2+}_{(aq)} \rightarrow Sn^{4+}_{(aq)} + Hg_2^{2+}_{(aq)}$$

(with oxidation numbers: Sn^{2+} as $+2$, Hg^{2+} as $+2$, Sn^{4+} as $+4$, Hg_2^{2+} as $+1$)

Oxidizing agent: Hg^{2+} Reducing agent: Sn^{2+}

Section 17.3 *Balancing Redox Equations: Oxidation Number Method*

19. Yes, the total electron loss by oxidation must equal the total electron gain by the reduction process.

21. (a)

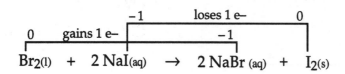

$$Br_{2(l)} + 2\,NaI_{(aq)} \rightarrow 2\,NaBr_{(aq)} + I_{2(s)}$$

(b)

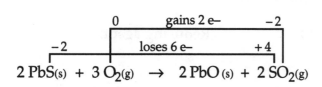

$$2\,PbS_{(s)} + 3\,O_{2(g)} \rightarrow 2\,PbO_{(s)} + 2\,SO_{2(g)}$$

23. (a)

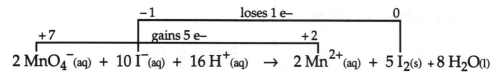

$$2\,MnO_4^{-}{}_{(aq)} + 10\,I^{-}{}_{(aq)} + 16\,H^{+}{}_{(aq)} \rightarrow 2\,Mn^{2+}{}_{(aq)} + 5\,I_{2(s)} + 8\,H_2O_{(l)}$$

(b)

$$Cu_{(s)} + 4\,H^{+}{}_{(aq)} + SO_4{}^{2-}{}_{(aq)} \rightarrow Cu^{2+}{}_{(aq)} + SO_{2(g)} + 2\,H_2O_{(l)}$$

Section 17.4 *Balancing Redox Equations: Half–Reaction Method*

25. (a) $SO_2(g) + 2\,H_2O(l) \rightarrow SO_4{}^{2-}(aq) + 4\,H^+(aq) + 2\,e^-$
    (b) $AsO_3{}^{3-}(aq) \rightarrow AsO_3{}^{-}(aq) + 2\,e^-$

27. (a) $ClO^-(aq) + H_2O(l) + 2\,e^- \rightarrow Cl^-(aq) + 2\,OH^-(aq)$
    (b) $MnO_4^-(aq) + 2\,H_2O(l) + 3\,e^- \rightarrow MnO_2(s) + 4\,OH^-(aq)$

29. (a) Oxidation: $3\,(Zn \rightarrow Zn^{2+} + 2\,e^-)$
        Reduction: $2\,(NO_3^- + 4\,H^+ + 3\,e^- \rightarrow NO + 2\,H_2O)$

        $3\,Zn(s) + 2\,NO_3^-(aq) + 8\,H^+(aq) \rightarrow 3\,Zn^{2+}(aq) + 2\,NO(g) + 4\,H_2O(l)$

    (b) Oxidation: $2\,(4\,H_2O + Mn^{2+} \rightarrow MnO_4^- + 8\,H^+ + 5\,e^-)$
        Reduction: $5\,(BiO_3^- + 6\,H^+ + 2\,e^- \rightarrow Bi^{3+} + 3\,H_2O)$

        $2\,Mn^{2+}(aq) + 5\,BiO_3^-(aq) + 14\,H^+(aq) \rightarrow 2\,MnO_4^-(aq) + 5\,Bi^{3+}(aq) + 7\,H_2O(l)$

31. (a) Oxidation: $3 (S^{2-} \rightarrow S + 2 e^-)$
Reduction: $2 (MnO_4^- + 2 H_2O + 3 e^- \rightarrow MnO_2 + 4 OH^-)$

$3 S^{2-}(aq) + 2 MnO_4^-(aq) + 4 H_2O(l) \rightarrow 3 S(s) + 2 MnO_2(s) + 8 OH^-(aq)$

(b) Oxidation: $Cu \rightarrow Cu^{2+} + 2 e^-$
Reduction: $ClO^- + H_2O + 2 e^- \rightarrow Cl^- + 2 OH^-$

$Cu(s) + ClO^-(aq) + H_2O(l) \rightarrow Cu^{2+}(aq) + Cl^-(aq) + 2 OH^-(aq)$

33. Oxidation: $Cl_2 + 2 H_2O \rightarrow 2 HOCl + 2 H^+ + 2 e^-$
Reduction: $Cl_2 + 2 e^- \rightarrow 2 Cl^-$

$Cl_2(g) + H_2O(l) \rightarrow Cl^-(aq) + HOCl(aq) + H^+(aq)$

35. Oxidation: $Cl_2 + 8 OH^- \rightarrow 2 ClO_2^- + 4 H_2O + 6 e^-$
Reduction: $3 (Cl_2 + 2 e^- \rightarrow 2 Cl^-)$

$4 Cl_2(g) + 8 OH^-(aq) \rightarrow 2 ClO_2^-(aq) + 6 Cl^-(aq) + 4 H_2O(l)$

Section 17.5 *Predicting Spontaneous Redox Reactions*

37.

Substances	Greater Tendency to be Reduced
(a) $Pb^{2+}(aq)$ or $Zn^{2+}(aq)$	$Pb^{2+}(aq)$
(b) $Fe^{3+}(aq)$ or $Al^{3+}(aq)$	$Fe^{3+}(aq)$
(c) $Ag^+(aq)$ or $I_2(s)$	$Ag^+(aq)$
(d) $Cu^{2+}(aq)$ or $Br_2(l)$	$Br_2(l)$

39.

Substances	Stronger Oxidizing Agent
(a) $F_2(g)$ or $Cl_2(g)$	$F_2(g)$
(b) $Ag^+(aq)$ or $Br_2(l)$	$Br_2(l)$
(c) $Cu^{2+}(aq)$ or $H^+(aq)$	$Cu^{2+}(aq)$
(d) $Mg^{2+}(aq)$ or $Mn^{2+}(aq)$	$Mn^{2+}(aq)$

41. (a) nonspontaneous   (b) spontaneous

43. (a) spontaneous   (b) spontaneous

45. $A(s) + B^+(aq) \rightarrow A^+(aq) + B(s)$
$B(s) + C^+(aq) \rightarrow B^+(aq) + C(s)$

From the first reaction, we can conclude that A has a greater tendency to undergo oxidation than B. From the second reaction, we can conclude that B has a greater tendency to undergo oxidation than C. Therefore, the higher *reduction potentials* are as follows: C > B > A.

47.

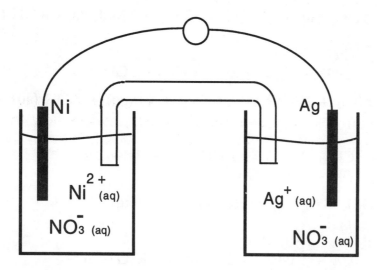

$$Ni(s) + 2\,AgNO_3(aq) \rightarrow 2\,Ag(s) + Ni(NO_3)_2(aq)$$

49.

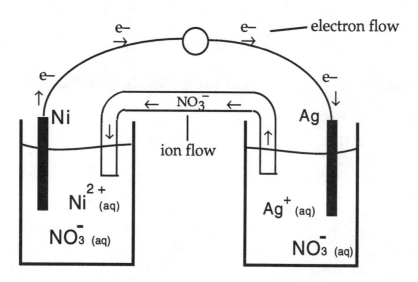

$$Ni(s) + 2\,AgNO_3(aq) \rightarrow 2\,Ag(s) + Ni(NO_3)_2(aq)$$

51. $Sn(s) + Cu^{2+}(aq) \rightarrow Cu(s) + Sn^{2+}(aq)$

(a) oxidation half-cell reaction: $Sn(s) \rightarrow Sn^{2+}(aq) + 2\,e^-$

(b) reduction half-cell reaction: $Cu^{2+}(aq) + 2\,e^- \rightarrow Cu(s)$

(c) anode: Sn electrode   cathode: Cu electrode

(d) direction of e^- flow: Sn anode $\rightarrow$ Cu cathode

(e) direction of SO_4^{2-} in the salt bridge: Cu half-cell $\rightarrow$ Sn half-cell

53. $Mg(s) + Mn(NO_3)_2(aq) \rightarrow Mn(s) + Mg(NO_3)_2(aq)$

(a) oxidation half-cell reaction: $Mg(s) \rightarrow Mg^{2+}(aq) + 2 e^-$
(b) reduction half-cell reaction: $Mn^{2+}(aq) + 2 e^- \rightarrow Mn(s)$
(c) anode: Mg electrode   cathode: Mn electrode
(d) direction of e^- flow: Mg anode $\rightarrow$ Mn cathode
(e) direction of NO_3^- in the salt bridge: Mn half-cell $\rightarrow$ Mg half-cell

(*Note:* If the anode is shown in the left half-cell, *electrons* flow from the left half-cell to the right half-cell. Conversely, *anions* flow through the salt bridge from the right half-cell to the left half-cell.)

Section 17.7 *Electrolytic Cells*

55.

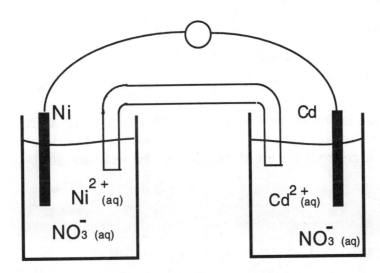

$$Ni(s) + Cd(NO_3)_2(aq) \rightarrow Cd(s) + Ni(NO_3)_2(aq)$$

57.

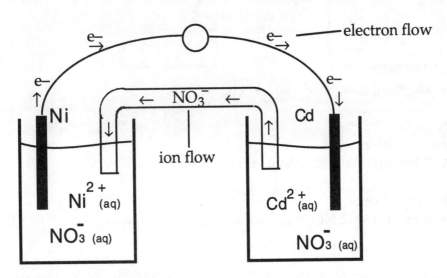

$$Ni(s) + Cd(NO_3)_2(aq) \rightarrow Cd(s) + Ni(NO_3)_2(aq)$$

59. $Ni(s) + Fe^{2+}(aq) \rightarrow Fe(s) + Ni^{2+}(aq)$

(a) oxidation half-cell reaction: $Ni(s) \rightarrow Ni^{2+}(aq) + 2\,e^-$

(b) reduction half-cell reaction: $Fe^{2+}(aq) + 2\,e^- \rightarrow Fe(s)$

(c) anode: Ni electrode   cathode: Fe electrode

(d) direction of e^- flow: Ni anode $\rightarrow$ Fe cathode

(e) direction of SO_4^{2-} in the salt bridge: Fe half-cell $\rightarrow$ Ni half-cell

61. $Cr(s) + Al(C_2H_3O_2)_3(aq) \rightarrow Al(s) + Cr(C_2H_3O_2)_3(aq)$

(a) oxidation half-cell reaction: $Cr(s) \rightarrow Cr^{3+}(aq) + 3\,e^-$

(b) reduction half-cell reaction: $Al^{3+}(aq) + 3\,e^- \rightarrow Al(s)$

(c) anode: Cr electrode   cathode: Al electrode

(d) direction of e^- flow: Cr anode $\rightarrow$ Al cathode

(e) direction of $C_2H_3O_2^-$ in the salt bridge: Al half-cell $\rightarrow$ Cr half-cell

(*Note:* If the anode is shown in the left half-cell, *electrons* flow from the left half-cell to the right half-cell. Conversely, *anions* flow through the salt bridge from the right half-cell to the left half-cell.)

General Exercises

63. $\underline{\text{Sodium Thiosulfate}}$ $\underline{\text{Oxidation Number of Sulfur}}$
    $Na_2S_2O_3$

    $2(\text{ox no Na}) + 2(\text{ox no S}) + 3(\text{ox no O}) = 0$
    $2(+1) + 2(\text{ox no S}) + 3(-2) = 0$
    $2(\text{ox no S}) + (-4) = 0$
    $2(\text{ox no S}) = +4$
    $\text{ox no S} = +2$

65. Net ionic equation:
    $Co(s) + Hg^{2+}(aq) \rightarrow Co^{2+}(aq) + Hg(l)$

67. Net ionic equation:
    $Zn(s) + 2\,H^+(aq) \rightarrow Zn^{2+}(aq) + H_2(g)$

69. Net ionic equation:
    $Cu(s) + 4\,H^+(aq) + 2\,NO_3^-(aq) \rightarrow Cu^{2+}(aq) + 2\,NO_2(g) + 2\,H_2O(l)$

Nuclear Chemistry

Section 18.1 *Natural Radioactivity*

1. <u>Principal Types of Natural Radiation</u>
   (1) alpha
   (2) beta
   (3) gamma

3. An alpha particle (α) is identical to a helium–4 nucleus.

5. Gamma rays (γ) are a form of radiant light energy.

Section 18.2 *Nuclear Equations*

7. | | Particle | Notation | | Particle | Notation |
   |---|---|---|---|---|---|
   | (a) | alpha particle | $^{4}_{2}\text{He}$ | (b) | beta particle | $^{0}_{-1}\text{e}$ |
   | (c) | gamma ray | $^{0}_{0}\gamma$ | (d) | positron | $^{0}_{+1}\text{e}$ |
   | (e) | neutron | $^{1}_{0}\text{n}$ | (f) | proton | $^{1}_{1}\text{H}$ |

9. (a) $^{175}_{78}\text{Pt} \rightarrow {}^{171}_{76}\text{Os} + {}^{4}_{2}\text{He}$

   (b) $^{28}_{13}\text{Al} \rightarrow {}^{28}_{14}\text{Si} + {}^{0}_{-1}\text{e}$

   (c) $^{55}_{27}\text{Co} \rightarrow {}^{55}_{26}\text{Fe} + {}^{0}_{+1}\text{e}$

   (d) $^{44}_{22}\text{Ti} + {}^{0}_{-1}\text{e} \rightarrow {}^{44}_{21}\text{Sc}$

11. (a) $^{221}_{88}X \rightarrow {}^{217}_{86}Rn + {}^{4}_{2}He$ $^{221}_{88}X = {}^{221}_{88}Ra$

(b) $^{43}_{19}X \rightarrow {}^{43}_{20}Ca + {}^{0}_{-1}e$ $^{43}_{19}X = {}^{43}_{19}K$

(c) $^{19}_{10}X \rightarrow {}^{19}_{9}F + {}^{0}_{+1}e$ $^{19}_{10}X = {}^{19}_{10}Ne$

(d) $^{37}_{18}X + {}^{0}_{-1}e \rightarrow {}^{37}_{17}Cl$ $^{37}_{18}X = {}^{37}_{18}Ar$

Section 18.3 *Radioactive Decay Series*

13. Let X = the parent nuclide

$$^{210}_{84}X \rightarrow {}^{206}_{82}Pb + {}^{4}_{2}He$$ The parent nuclide is: $^{210}_{84}Po$.

15. $^{207}_{81}X \rightarrow {}^{207}_{82}Pb + {}^{0}_{-1}e$ The parent nuclide is: $^{207}_{81}Tl$.

17. $^{212}_{84}X \rightarrow {}^{208}_{82}Pb + {}^{4}_{2}He$ The parent nuclide is: $^{212}_{84}Po$.

19. <u>Decay Series for Thorium–232</u>   <u>Answers</u>

$^{232}_{90}Th \xrightarrow{\alpha} {}^{228}_{88}Ra \xrightarrow{\beta} {}^{228}_{89}Ac$ (a) α (b) β

$\downarrow \beta$ (c) β (d) α

$^{220}_{86}Rn \xleftarrow{\alpha} {}^{224}_{88}Ra \xleftarrow{\alpha} {}^{228}_{90}Th$ (e) α (f) α

$\alpha\downarrow$ (g) α (h) β

$^{216}_{84}Po \xrightarrow{\alpha} {}^{212}_{82}Pb \xrightarrow{\beta} {}^{212}_{83}Bi$ (i) β (j) α

$\downarrow \beta$

$^{208}_{82}Pb \xleftarrow{\alpha} {}^{212}_{84}Po$

Section 18.4 *Radioactive Half–Life*

21. Half of 100% of a radionuclide is 50%. Half of 50% of a radionuclide is 25%. Thus, it takes two half-lives to reach 25% of an original radionuclide sample from the complete sample.

23. $15\ t_{1/2} \times \dfrac{24{,}400\ \text{years}}{1\ t_{1/2}} = 366{,}000\ \text{years}$

25.

Activity	Elapsed Time
100%	$0\ t_{1/2}$
50%	$1\ t_{1/2}$
25%	$2\ t_{1/2}$
12.5%	$3\ t_{1/2}$
6.25%	$4\ t_{1/2}$

$4\ t_{1/2} \times \dfrac{5730\ \text{years}}{1\ t_{1/2}} \approx 22{,}900\ \text{years old}$

27. $60\ \cancel{\text{hours}} \times \dfrac{1\ t_{1/2}}{15\ \cancel{\text{hours}}} = 4\ t_{1/2}$

$80\ \text{mg Na–24} \times \frac{1}{2} \times \frac{1}{2} \times \frac{1}{2} \times \frac{1}{2} = 5\ \text{mg Na–24}$

29.

Activity	Elapsed Time
560 dpm	$0\ t_{1/2}$
280 dpm	$1\ t_{1/2}$
140 dpm	$2\ t_{1/2}$
70 dpm	$3\ t_{1/2}$
35 dpm	$4\ t_{1/2}$

$4\ t_{1/2} \times \dfrac{74\ \text{days}}{1\ t_{1/2}} = 296\ \text{days}$

31.

Activity	Elapsed Time
1200 dpm	$0\ t_{1/2}$
600 dpm	$1\ t_{1/2}$
300 dpm	$2\ t_{1/2}$
150 dpm	$3\ t_{1/2}$
75 dpm	$4\ t_{1/2}$

$\text{half-life} = \dfrac{21.2\ \text{years}}{4\ t_{1/2}} = 5.30\ \text{years/half-life}$

Section 18.5 *Radionuclide Applications*

33. Fossils up to 50,000 years old can be dated using radiocarbon dating (that is, carbon–14).

35. Uranium–lead dating can be used to estimate geological events occurring billions of years ago.

37. Plutonium–238 releases high-energy gamma rays that are used to power heart pacemakers.

39. The β–emitting radionuclide iodine–131 can be used to measure the activity of the thyroid gland.

41. The γ–emitting radionuclide technetium–99 can be used to diagnose and locate brain tumors.

Section 18.6 *Artificial Radioactivity*

43. $^{23}_{11}\text{Na} + ^{1}_{1}\text{H} \rightarrow ^{23}_{12}\text{Mg} + ^{1}_{0}\text{n}$ The radionuclide is: $^{23}_{12}\text{Mg}$.

45. $^{6}_{3}\text{Li} + ^{1}_{0}\text{n} \rightarrow ^{3}_{1}\text{H} + ^{4}_{2}\text{He}$ The radionuclide is: $^{3}_{1}\text{H}$.

47. $^{59}_{27}\text{Co} + ^{1}_{0}\text{n} \rightarrow ^{56}_{25}\text{Mn} + ^{4}_{2}\text{He}$ The target nuclide is: $^{59}_{27}\text{Co}$.

49. $^{10}_{5}\text{B} + ^{4}_{2}\text{He} \rightarrow ^{14}_{7}\text{N} + ^{0}_{0}\gamma$ The projectile particle is: $^{4}_{2}\text{He}$.

51. $^{243}_{95}\text{Am} + ^{22}_{10}\text{Ne} \rightarrow ^{260}_{105}\text{X} + 5\,^{1}_{0}\text{n}$ The Russians created: $^{260}_{105}\text{X}$.

53. $^{54}_{24}\text{Cr} + ^{207}_{82}\text{Pb} \rightarrow ^{260}_{106}\text{X} + ^{1}_{0}\text{n}$ The Russians created: $^{260}_{106}\text{X}$.

Section 18.7 *Nuclear Fission*

55. $^{1}_{0}\text{n} \xrightarrow{\text{1st}} 2\,^{1}_{0}\text{n} \xrightarrow{\text{2nd}} 4\,^{1}_{0}\text{n} \xrightarrow{\text{3rd}} 8\,^{1}_{0}\text{n}$

After the third step, 8 neutrons are produced.

57. Uranium–235 can fission in many different ways to give various products. The number of neutrons released from a single fission is usually 2 or 3, but the *average* number of neutrons is 2.4.

59. $^{235}_{92}U + ^{1}_{0}n \rightarrow ^{144}_{54}Xe + ^{90}_{38}Sr + 2^{1}_{0}n$ The number of neutrons is 2.

61. $^{239}_{94}Pu + ^{1}_{0}n \rightarrow ^{137}_{55}Cs + ^{100}_{39}Y + 3^{1}_{0}n$ The number of neutrons is 3.

63. $^{233}_{92}U + ^{1}_{0}n \rightarrow ^{143}_{56}Ba + ^{88}_{36}Kr + 3^{1}_{0}n$ The fissionable nuclide is: $^{233}_{92}U$.

Section 18.8 *Mass Defect and Binding Energy*

65. Tritium has one proton and two neutrons:

$$1(1.0073 \text{ amu}) + 2(1.0087 \text{ amu}) = 3.0247 \text{ amu}$$

Mass defect: $3.0247 \text{ amu} - 3.0155 \text{ amu} = 0.0092 \text{ amu}$

67. Lithium–7 has three protons and four neutrons:

$$3(1.0073 \text{ amu}) + 4(1.0087 \text{ amu}) = 7.0567 \text{ amu}$$

Mass defect: $7.0567 \text{ amu} - 7.0160 \text{ amu} = 0.0407 \text{ amu}$

69. Boron–11 has five protons and six neutrons:

$$5(1.0073 \text{ amu}) + 6(1.0087 \text{ amu}) = 11.0887 \text{ amu}$$

Mass defect: $11.0887 \text{ amu} - 11.0093 \text{ amu} = 0.0794 \text{ amu}$

71. Binding energy: $E = mc^2$

$$E = \frac{0.0084 \text{ g}}{1 \text{ mol}} \times \frac{1 \text{ kg}}{1000 \text{ g}} \times \left(\frac{3.00 \times 10^8 \text{ m}}{s}\right)^2 = 7.6 \times 10^{11} \text{ J/mol}$$

Section 18.9 *Nuclear Fusion*

73. $^{2}_{1}H + ^{2}_{1}H \rightarrow ^{3}_{1}H + ^{0}_{+1}e + ^{1}_{0}X$ The particle X is: $^{1}_{0}n$.

75. $^{3}_{2}He + ^{1}_{1}X \rightarrow ^{4}_{2}He + ^{0}_{+1}e$ The particle X is: $^{1}_{1}H$.

77. $^{1}_{1}X + ^{1}_{1}X \rightarrow ^{2}_{1}H + ^{0}_{+1}e$ The nuclide X is: $^{1}_{1}H$.

General Exercises

79. $^{241}_{94}\text{Pu} \rightarrow {}^{4}_{2}\text{He} + {}^{237}_{92}\text{X}$ The daughter product is: $^{237}_{92}\text{U}$.

81. $^{137}_{53}\text{X} \rightarrow {}^{136}_{54}\text{Xe} + {}^{0}_{-1}\text{e} + {}^{1}_{0}\text{n}$ The radionuclide X is: $^{137}_{53}\text{I}$.

83. Since an activity of ~ 7 dpm per gram indicates that one half-life has expired, the age of Crater Lake is ~ 6000 years because $1\ t_{1/2} = 5730$ years.

85. $^{241}_{95}\text{X} \rightarrow {}^{237}_{93}\text{Np} + {}^{4}_{2}\text{He}$ The radionuclide is: $^{241}_{95}\text{Am}$.

87. $^{238}_{92}\text{U} + 15\ {}^{1}_{0}\text{n} \rightarrow {}^{253}_{99}\text{Es} + 7\ {}^{0}_{-1}\text{e}$ The number of neutrons is 15.

89. $^{253}_{99}\text{Es} + {}^{4}_{2}\text{He} \rightarrow {}^{1}_{0}\text{n} + {}^{256}_{101}\text{X}$ The nuclide X is: $^{256}_{101}\text{Md}$.

91. $^{6}_{3}\text{Li} + {}^{1}_{0}\text{n} \rightarrow {}^{3}_{1}\text{H} + {}^{4}_{2}\text{X} + \text{energy}$ The particle X is: $^{4}_{2}\text{He}$.

Organic Chemistry

Section 19.1 *Hydrocarbons*

1. Approximately 90% of all chemical compounds contain carbon and are considered organic.

3. The primary source of hydrocarbons is petroleum (crude oil).

5.
Hydrocarbon	Classification
(a) alkanes	saturated hydrocarbon
(b) alkenes	unsaturated hydrocarbon
(c) alkynes	unsaturated hydrocarbon

Section 19.2 *Alkanes*

7.
   Molecular Formula — **Class of Compound**

   (a) $C_{10}H_{22}$

   The molecular formula of the alkanes is C_nH_{2n+2}. Since $C_{10}H_{22}$ follows the general formula (n = 10), it is an alkane.

   (b) $C_{14}H_{28}$

   The molecular formula of the alkanes is C_nH_{2n+2}. Since $C_{14}H_{28}$ does not fit the formula, it is *not* an alkane.

9.
   Structural Formula — **IUPAC Name**

   (a) $CH_3-CH_2-CH_2-CH_3$ — butane
   (b) $CH_3-CH_2-CH_2-CH_2-CH_2-CH_3$ — hexane
   (c) $CH_3-CH_2-CH_2-CH_2-CH_2-CH_2-CH_2-CH_3$ — octane
   (d) $CH_3-CH_2-CH_2-CH_2-CH_2-CH_2-CH_2-CH_2-CH_2-CH_3$ — decane

11. <u>Five Isomers of Hexane</u> (C_6H_{14})

$CH_3 - CH_2 - CH_2 - CH_2 - CH_2 - CH_3$

$$CH_3 - \underset{\underset{CH_3}{|}}{CH} - CH_2 - CH_2 - CH_3$$

$$CH_3 - CH_2 - \underset{\underset{CH_3}{|}}{CH} - CH_2 - CH_3$$

$$CH_3 - \underset{\underset{CH_3}{|}}{CH} - \underset{\underset{CH_3}{|}}{CH} - CH_3$$

$$CH_3 - \underset{\underset{CH_3}{\overset{CH_3}{|}}}{\overset{|}{C}} - CH_2 - CH_3$$

13. <u>Two Isomers of Bromopropane</u> (C_3H_7Br)

$CH_3–CH_2–CH_2–Br$ $\qquad\qquad$ $CH_3–CH(Br)–CH_3$

15. <u>Alkyl Substituents</u> $\qquad\qquad$ <u>IUPAC Name</u>

   (a) $CH_3–$ $\qquad\qquad$ methyl
   (b) $CH_3CH_2–$ $\qquad\qquad$ ethyl

17. (a) 2–methylpentane $\qquad$ (b) 2–methyl–4–ethylhexane
   (c) 2,4,4–trimethylheptane $\qquad$ (d) 3,4,5–trimethyloctane

19. (a) $CH_4 + 2\,O_2 \rightarrow CO_2 + 2\,H_2O$
   (b) $C_5H_{12} + 8\,O_2 \rightarrow 5\,CO_2 + 6\,H_2O$
   (c) $C_3H_8 + 5\,O_2 \rightarrow 3\,CO_2 + 4\,H_2O$
   (d) $C_7H_{16} + 11\,O_2 \rightarrow 7\,CO_2 + 8\,H_2O$

Section 19.3 *Alkenes and Alkynes*

21.   <u>Molecular Formula</u> $\qquad\qquad$ <u>Class of Compound</u>

   (a) $C_{10}H_{22}$ $\qquad\qquad$ The molecular formula of the alkenes is C_nH_{2n}. Since $C_{10}H_{22}$ does not fit the formula, it is *not* an alkene.

   (b) $C_{14}H_{28}$ $\qquad\qquad$ The molecular formula of the alkenes is C_nH_{2n}. Since $C_{14}H_{28}$ follows the general formula (n = 14), it is an alkene.

23.   <u>Structural Formula</u>   <u>IUPAC Name</u>

(a) $CH_2=CH-CH_2-CH_3$ 1–butene
(b) $CH_2=CH-CH_2-CH_2-CH_3$ 1–pentene
(c) $CH_3-CH=CH-CH_2-CH_2-CH_3$ 2–hexene
(d) $CH_3-CH_2-CH_2-CH=CH-CH_2-CH_2-CH_3$ 4–octene

25.   <u>Molecular Formula</u>   <u>Class of Compound</u>

(a) $C_{10}H_{18}$ The molecular formula of the alkynes is C_nH_{2n-2}. Since $C_{10}H_{18}$ follows the general formula (n = 10), it is an alkyne.

(b) $C_{14}H_{24}$ The molecular formula of the alkynes is C_nH_{2n-2}. Since $C_{14}H_{24}$ does not fit the formula, it is *not* an alkyne.

27.   <u>Structural Formula</u>   <u>IUPAC Name</u>

(a) $CH_3-C\equiv C-CH_2-CH_3$ 2–pentyne
(b) $CH\equiv C-CH_2-CH_2-CH_3$ 1–pentyne
(c) $CH_3-CH_2-C\equiv C-CH_2-CH_3$ 3–hexyne
(d) $CH_3-C\equiv C-CH_2-CH_2-CH_2-CH_3$ 2–heptyne

29. <u>Two Isomers of Straight-Chain Pentene</u> (C_5H_{10})

$CH_2=CH-CH_2-CH_2-CH_3$ $CH_3-CH=CH-CH_2-CH_3$

31. (a)

$$CH_3-CH=CH-\overset{\overset{\displaystyle CH_3}{|}}{CH}-CH_3$$

4–methyl–2–pentene

(b)

$$CH_3-\overset{\overset{\displaystyle CH_3}{|}}{CH}-CH_2-\overset{\overset{\displaystyle CH_3}{|}}{C}=CH-CH_3$$

3,5–dimethyl–2–hexene

33. (a)

$$CH_3-C\equiv C-\overset{\overset{\displaystyle CH_3}{|}}{\underset{\underset{\displaystyle CH_3}{|}}{C}}-CH_2-CH_2-CH_3$$

4,4–dimethyl–2–heptyne

(b)

$$CH_3-CH_2-CH_2-\overset{}{\underset{\underset{\displaystyle CH_3}{|}}{CH}}-\overset{\overset{\displaystyle CH_3}{|}}{CH}-\overset{}{\underset{\underset{\displaystyle CH_3}{|}}{CH}}-C\equiv CH$$

3,4,5–trimethyl–1–octyne

35. (a) $CH_2{=}CH_2 + 3\,O_2 \rightarrow 2\,CO_2 + 2\,H_2O$

(b) $CH_3{-}CH{=}CH_2 + H_2 \rightarrow CH_3{-}CH_2{-}CH_3$

(c) $CH_3{-}CH{=}CH{-}CH_3 + Br_2 \rightarrow CH_3{-}CHBr{-}CHBr{-}CH_3$

Section 19.4 *Aromatic Hydrocarbons*

37. <u>Kekulé Structures of Benzene</u> (C_6H_6)

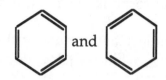

39. <u>Three Isomers of Xylene</u> $[C_6H_4(CH_3)_2]$

CH₃ structures labelled *ortho*, *meta*, *para*

Section 19.5 *Hydrocarbon Derivatives*

41.
General Formula	Class of Compound
(a) R—O—R′	ether
(b) R—X	organic halide
(c) Ar—OH	phenol
(d) R—NH₂	amine

43.
Chemical Formula	Class of Compound
(a) $CH_3{-}NH_2$	amine
(b) $CH_3CH_2{-}F$	organic halide

(c) 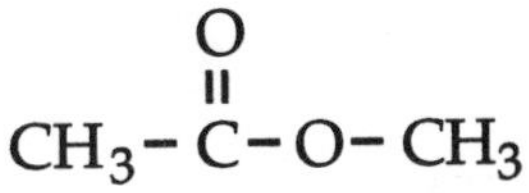 $CH_3-\overset{\overset{\displaystyle O}{\|}}{C}-O-CH_3$ ester

(d) 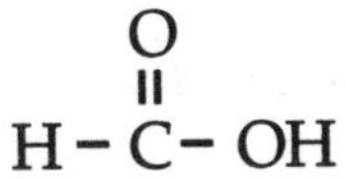 $H-\overset{\overset{\displaystyle O}{\|}}{C}-OH$ carboxylic acid

(e) 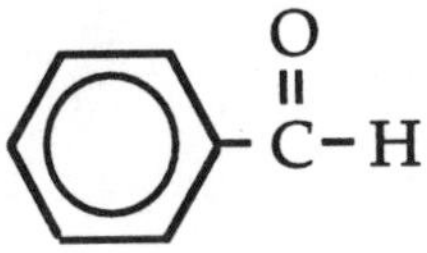 aldehyde

(f) 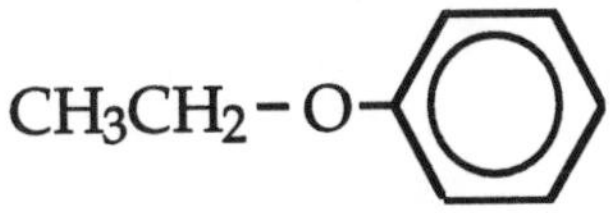 CH_3CH_2-O- ether

(g) 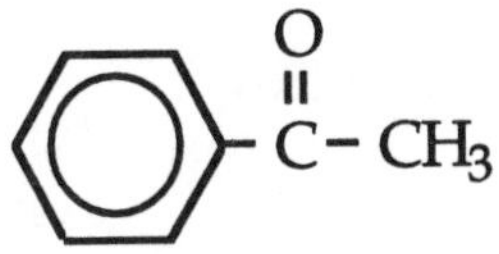 ketone

(h) 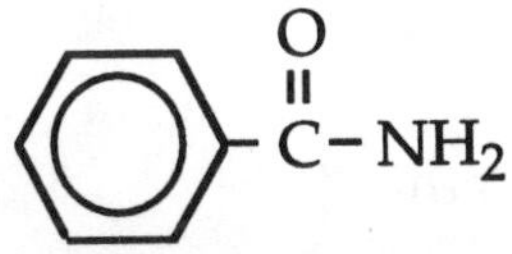 amide

Section 19.6 *Organic Halides*

45.   <u>Organic Halide</u>   <u>IUPAC Name</u>   <u>Common Name</u>

	Organic Halide	IUPAC Name	Common Name
(a)	CH_3-I	iodomethane	"methyl iodide"
(b)	CH_3-CH_2-Br	bromoethane	"ethyl bromide"
(c)	$CH_3-CH_2-CH_2-F$	1–fluoropropane	"propyl fluoride"
(d)	$(CH_3)_2-CH-Cl$	2–chloropropane	"isopropyl chloride"

47.   1,1,1–trichloroethane CH_3-CCl_3

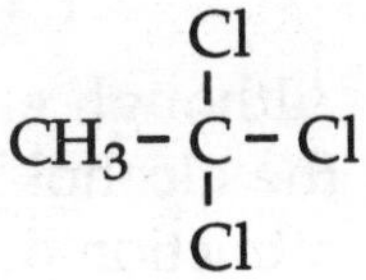

$$CH_3-\overset{\overset{\displaystyle Cl}{|}}{\underset{\underset{\displaystyle Cl}{|}}{C}}-Cl$$

49.   <u>Solvent</u>   <u>Solubility of Organic Halides</u>

(a) water   insoluble in water

(b) hydrocarbons   soluble in hydrocarbons

Section 19.7 *Alcohols, Phenols, and Ethers*

51.   <u>Alcohol</u>   <u>IUPAC Name</u>

    (a) $CH_3–CH_2–CH_2–CH_2–OH$ 1–butanol
    (b) $CH_3–CH_2–CH(OH)–CH_3$ 2–butanol
    (c) $CH_2(OH)–CH_2–CH_2–CH_3$ 1–butanol
    (d) $CH_3–CH(OH)–CH_2–CH_3$ 2–butanol

53.   <u>Phenol</u>   <u>IUPAC Name</u>

    (a)   ⬡–OH   phenol

    (b) CH_3–⬡–OH   *para*–methylphenol

55.   <u>Ether</u>   <u>Common Name</u>

    (a) $CH_3–O–CH_3$ "dimethyl ether"

    (b) $CH_3–O–CH_2CH_3$ "methyl ethyl ether"

    (c) $CH_3CH_2CH_2–O–CH_2CH_2CH_3$ "dipropyl ether"

    (d)   ⬡–O–CH_2CH_3 "phenyl ethyl ether"

57.   Although ethyl alcohol and dimethyl ether have the same molecular mass, the alcohol has a higher boiling point. Alcohols have strong intermolecular attraction due to hydrogen bonding, while ethers are attracted by weaker dipole forces.

59.   <u>Alcohol</u> (C_2H_6O) <u>Ether</u> (C_2H_6O)

    $CH_3 - CH_2 - OH$ $CH_3 - O - CH_3$

Section 19.8 *Amines*

61. | Amine | IUPAC Name | Common Name |
|---|---|---|
| (a) $CH_3-CH_2-NH_2$ | amino ethane | "ethyl amine" |
| (b) $CH_3-CH_2-CH_2-NH_2$ | 1–amino propane | "propyl amine" |

63. Although ethyl amine and propane have about the same molecular mass, the amine has a higher boiling point. Amines have strong intermolecular attraction due to hydrogen bonding, while alkanes are attracted by weaker dispersion forces.

Section 19.9 *Aldehydes and Ketones*

65. | Aldehyde | IUPAC Name | Common Name |
|---|---|---|
| (a) $H-\overset{\overset{O}{\|\|}}{C}-H$ | methanal | "formaldehyde" |
| (b) $CH_3-\overset{\overset{O}{\|\|}}{C}-H$ | ethanal | "acetaldehyde" |
| (c) $CH_3CH_2-\overset{\overset{O}{\|\|}}{C}-H$ | propanal | "propionaldehyde" |
| (d) $CH_3CH_2CH_2-\overset{\overset{O}{\|\|}}{C}-H$ | butanal | "butyraldehyde" |

67. | Ketone | IUPAC Name | Common Name |
|---|---|---|
| (a) $CH_3-\overset{\overset{O}{\|\|}}{C}-CH_3$ | 2–propanone | "dimethyl ketone" |
| (b) $CH_3-\overset{\overset{O}{\|\|}}{C}-CH_2CH_3$ | 2–butanone | "methyl ethyl ketone" |
| (c) $CH_3-\overset{\overset{O}{\|\|}}{C}-CH_2CH_2CH_3$ | 2–pentanone | "methyl propyl ketone" |
| (d) $CH_3CH_2-\overset{\overset{O}{\|\|}}{C}-CH_2CH_3$ | 3–pentanone | "diethyl ketone" |

69. (a) Aldehydes generally have *higher boiling points* than hydrocarbons with similar molecular mass. Aldehydes demonstrate a strong intermolecular attraction due to dipole forces, while hydrocarbons are attracted by weaker dispersion forces.

(b) Ketones generally have *higher boiling points* than hydrocarbons with similar molecular mass. Ketones demonstrate a strong intermolecular attraction due to dipole forces, while hydrocarbons are attracted by weaker dispersion forces.

71. Aldehyde (C_3H_6O)   Ketone (C_3H_6O)

$$CH_3CH_2-\overset{\overset{\displaystyle O}{\|}}{C}-H \qquad\qquad CH_3-\overset{\overset{\displaystyle O}{\|}}{C}-CH_3$$

Section 19.10 *Carboxylic Acids, Esters, and Amides*

73.   Carboxylic Acid   IUPAC Name   Common Name

(a) $H-\overset{\overset{\displaystyle O}{\|}}{C}-OH$ methanoic acid   "formic acid"

(b) $CH_3-\overset{\overset{\displaystyle O}{\|}}{C}-OH$ ethanoic acid   "acetic acid"

(c) $CH_3CH_2-\overset{\overset{\displaystyle O}{\|}}{C}-OH$ propanoic acid   "propionic acid"

(d) $CH_3CH_2CH_2-\overset{\overset{\displaystyle O}{\|}}{C}-OH$ butanoic acid   "butyric acid"

75.   Ester   IUPAC Name   Common Name

(a) $H-\overset{\overset{\displaystyle O}{\|}}{C}-O-CH_2CH_3$ ethyl methanoate "ethyl formate"

(b) $CH_3-\overset{\overset{\displaystyle O}{\|}}{C}-O-CH_3$ methyl ethanoate "methyl acetate"

(c) $CH_3CH_2-\overset{\overset{\displaystyle O}{\|}}{C}-O-CH_2CH_3$ ethyl propanoate   "ethyl propionate"

(d) $H-\overset{\overset{\displaystyle O}{\|}}{C}-O-$ ⬡   phenyl methanoate "phenyl formate"

77.   <u>Amide</u>   <u>IUPAC Name</u>   <u>Common Name</u>

(a) $H-\overset{\overset{\displaystyle O}{\|}}{C}-NH_2$ methanamide   "formamide"

(b) ⬡$-\overset{\overset{\displaystyle O}{\|}}{C}-NH_2$ benzamide   —

79.   (a)   Carboxylic acids have an O–H group and thus can *hydrogen bond*.
      (b)   Esters do not have an O–H or an N–H group and cannot *hydrogen bond*.
      (c)   Amides have an N–H group and thus can *hydrogen bond*.

81.   Although propionic acid and methyl acetate have the same molecular mass, propionic acid has a *higher boiling point*. Acids have strong intermolecular attraction due to hydrogen bonding, while esters are attracted by weaker dipole forces.

83.   (a) $CH_3-\overset{\overset{\displaystyle O}{\|}}{C}-OH \ + \ CH_3-OH \quad \overset{H_2SO_4}{\longrightarrow} \quad CH_3-\overset{\overset{\displaystyle O}{\|}}{C}-O-CH_3 \ + \ H_2O$

      (b) ⬡$-\overset{\overset{\displaystyle O}{\|}}{C}-OH \ + \ NH_3 \quad \overset{\Delta}{\longrightarrow} \quad$ ⬡$-\overset{\overset{\displaystyle O}{\|}}{C}-NH_2 \ + \ H_2O$

85.   The ester produced from the reaction of phenol and acetic acid is phenyl ethanoate ("phenyl acetate").

87.   The amide produced from the reaction of ammonia and formic acid is methanamide ("formamide").

General Exercises

89.   Incorrect Name   IUPAC Name

  (a) ethylmethane   propane
  (b) propylethane   pentane

91.   Compound   Classification

  (a) eicosane   saturated (–ane suffix)
  (b) dodecene   unsaturated (–ene suffix)

93.   Compounds   Higher Boiling Point

  (a) CH_3-O-CH_3 or CH_3CH_2-OH CH_3CH_2-OH
  (b) $CH_3CH_2CH_2-NH_2$ or $CH_3CH_2CH_2-F$ $CH_3CH_2CH_2-NH_2$

  In each of the above pairs, the compound with the higher boiling point
  has an –OH or $-NH_2$ group, which can hydrogen bond. Intermolecular
  hydrogen bonds increase the effective molar mass of the compound and
  therefore the compound requires more energy—a higher temperature—
  to boil and escape the liquid state.

95.   Compound   Class of Compound

  (a) **chlor**dane   organic halide
  (b) cortis**one**   ketone
  (c) acetylsalicy**lic acid**   carboxylic acid
  (d) sulfanil**amide**   amide
  (e) nicot**ine**   amine

Biochemistry

Section 20.1 *Biological Compounds*

1. (a) A *protein* is a polymer compound consisting of amino acids.
   (b) A *nucleic acid* is a polymer compound consisting of sugar molecules with an attached organic base, and phosphoric acid.

3. (a) The repeating units in a protein are joined by *peptide linkages*.
   (b) The repeating units in a nucleic acid are joined by *phosphate linkages*.

5. <u>Biological Compound</u>   <u>Organic Functional Group</u>

   (a) amino acid   carboxylic acid, amine
   (b) sugar   aldehyde, ketone, alcohol
   (c) fatty acid   carboxylic acid
   (d) organic base   amine

Section 20.2 *Proteins*

7. A covalent amide bond is responsible for linking two amino acids together in the primary structure of a protein.

9. The primary structure of a protein refers to the amino acid sequence. The secondary structure of a protein refers to the shape of the chain, for example, an α–helix or a pleated sheet.

11. The dipeptide formed from two molecules of alanine is named alanylalanine (ala–ala):

$$H_2N-CH-\overset{\overset{\textstyle O}{\|}}{C}-NH-CH-\overset{\overset{\textstyle O}{\|}}{C}-OH$$

with CH_3 groups below the two CH carbons.

13. The dipeptide formed from a molecule of serine and a molecule of tyrosine is named seryltyrosine (ser–tyr):

$$H_2N-CH-\overset{\overset{\displaystyle O}{\|}}{C}-NH-CH-\overset{\overset{\displaystyle O}{\|}}{C}-OH$$

with side chains CH_2-OH on the first carbon and CH_2 attached to a phenol ring (OH) on the second carbon.

15.

Tripeptide	Amino Acid Sequence
1	arg–his–met
2	arg–met–his
3	his–arg–met
4	his–met–arg
5	met–arg–his
6	met–his–arg

17. The two amino acids in NutraSweet are *aspartic acid* and the methyl ester of *phenyl alanine*.

19. Two of the amino acids in oxytocin differ from those in vasopressin. In oxytocin, *isoleucine* replaces phenyl alanine, and *leucine* replaces arginine.

Section 20.3 *Enzymes*

21. In Step 1 of enzyme catalysis, the substrate molecule binds to the active site on the enzyme.

23. In the lock–and–key model, the "teeth" on the key represent an active site on the enzyme molecule.

25. The term for a molecule that blocks an enzyme and prevents a substrate reaction is called an *inhibitor*.

27. Three defining characteristics of an enzyme are as follows.
    (1) An enzyme is selective for a given molecule and reacts to give specific products.
    (2) An enzyme can accelerate a biochemical reaction by a factor of a million or more.
    (3) Only a trace amount of enzyme is necessary to catalyze a biochemical reaction.

Section 20.4 *Carbohydrates*

29. An aldose sugar has an *aldehyde* functional group as well as one or more *alcohol* groups.

31. A pentose sugar molecule contains five carbon atoms.

33. A monosaccharide is a simple sugar molecule such as glucose, whereas a disaccharide has two simple sugar molecules joined by a glycoside linkage.

35. See textbook Figure 20.9.

37. See textbook Figure 20.10.

39. $C_{12}H_{22}O_{11}$ + H_2O $\xrightarrow{H^+}$ $C_6H_{12}O_6$ + $C_6H_{12}O_6$

 maltose + water $\rightarrow$ glucose + glucose

41. Glycogen is a small polysaccharide composed of glucose sugar molecules as the repeating monosaccharide.

Section 20.5 *Lipids*

43.

Characteristic	Lipid
(a) animal source	fat
(b) plant source	oil
(c) semisolid	fat
(d) insoluble in water	fat and oil

45. The structural formula for the triglyceride of stearic acid is:

$$CH_2-O-\overset{\overset{\displaystyle O}{\|}}{C}-(CH_2)_{16}-CH_3$$

$$CH-O-\overset{\overset{\displaystyle O}{\|}}{C}-(CH_2)_{16}-CH_3$$

$$CH_2-O-\overset{\overset{\displaystyle O}{\|}}{C}-(CH_2)_{16}-CH_3$$

Stearic acid: $CH_3-(CH_2)_{16}-COOH$

47.

Name of Fatty Acid	Structure of Fatty Acid
(a) palmitic acid	$HOOC–(CH_2)_{14}–CH_3$
(b) lauric acid	$HOOC–(CH_2)_{10}–CH_3$
(c) myristic acid	$HOOC–(CH_2)_{12}–CH_3$

49. The saponification of carnauba wax with aqueous NaOH produces a fatty acid and long-chain alcohol.

$$CH_3–(CH_2)_{24}–\overset{\overset{\displaystyle O}{\|}}{C}—O—(CH_2)_{29}–CH_3 \ + \ NaOH \ \rightarrow$$

$$CH_3–(CH_2)_{24}–COO^-Na^+ \ + \ CH_3–(CH_2)_{29}–OH$$

51. The water–insoluble vitamins (A, D, E, K) are considered lipids.

Section 20.6 *Nucleic Acids*

53. The three general components of a DNA nucleotide are: deoxyribose sugar, an organic base, and phosphoric acid.

55. The four nitrogen bases in DNA are: adenine (A), cytosine (C), guanine (G), and thymine (T).

57. The structures of deoxyribose (in DNA) and ribose (in RNA) are identical except that deoxyribose is missing a hydroxyl (–OH) group.

59. A DNA molecule has two polymer strands of nucleotides in the shape of a double helix.

61. Hydrogen bonds between organic bases hold together the two strands of nucleotides in the DNA double helix.

63. During DNA replication, an adenine (A) base on the template DNA strand will code for a thymine (T) base on the complementary DNA strand.

65. During RNA transcription, an adenine (A) base on the template DNA strand will code for a uracil (U) base on the growing RNA strand.

General Exercises

67. A protein that provides strength, such as in muscle tissue, has a long and extended shape.

69. When a protein undergoes denaturation in aqueous acid, hydrogen bonds are broken and the protein loses its secondary and tertiary structure and acquires a random shape.

71. Similar to hair, skin, and nails in humans, the main component in animal horns, hooves, claws, and feathers is protein.

73. There are three possible trinucleotide sequences for two different nucleotides.

Trinucleotide	Nucleotide Sequence
1	A–C–C
2	C–C–A
3	C–A–C

75. A DNA molecule replicates by first unwinding and breaking hydrogen bonds between organic bases in the double helix. Second, each single strand of DNA acts as a template to synthesize a complementary strand of DNA one nucleotide at a time.

Since organic bases always pair in the same way, $A = T$ or $C \equiv G$, each new nucleotide in a growing strand of DNA must complement the existing nucleotide in the template strand. If the nucleotide in the template strand contains adenine (A), it will code for thymine (T) on the growing strand. If the nucleotide contains cytosine (C), it will code for guanine (G). After the synthesis is complete, the template and complementary strands are joined by hydrogen bonds and the structure is identical to the original double helix.

77. There are 24 tetrapeptides containing four different amino acids.

Tetrapeptide	Amino Acid Sequence	Tetrapeptide	Amino Acid Sequence
1	gly–ala–ser–trp	13	ser–trp–gly–ala
2	gly–ala–trp–ser	14	ser–trp–ala–gly
3	gly–ser–trp–ala	15	ser–gly–ala–trp
4	gly–ser–ala–trp	16	ser–gly–trp–ala
5	gly–trp–ala–ser	17	ser–ala–trp–gly
6	gly–trp–ser–ala	18	ser–ala–gly–trp
7	ala–ser–trp–gly	19	trp–gly–ala–ser
8	ala–ser–gly–trp	20	trp–gly–ser–ala
9	ala–trp–gly–ser	21	trp–ala–ser–gly
10	ala–trp–ser–gly	22	trp–ala–gly–ser
11	ala–gly–ser–trp	23	trp–ser–gly–ala
12	ala–gly–trp–ser	24	trp–ser–ala–gly